Coastal Research Library

Volume 14

Series Editor

Charles W. Finkl
Department of Geosciences
Florida Atlantic University
Boca Raton, FL
USA

The aim of this book series is to disseminate information to the coastal research community. The Series covers all aspects of coastal research including but not limited to relevant aspects of geological sciences, biology (incl. ecology and coastal marine ecosystems), geomorphology (physical geography), climate, littoral oceanography, coastal hydraulics, environmental (resource) management, engineering, and remote sensing. Policy, coastal law, and relevant issues such as conflict resolution and risk management would also be covered by the Series. The scope of the Series is broad and with a unique cross-disciplinary nature. The Series would tend to focus on topics that are of current interest and which carry some import as opposed to traditional titles that are esoteric and non-controversial. Monographs as well as contributed volumes are welcomed.

More information about this series at http://www.springer.com/series/8795

V. Santiago-Fandiño • H. Tanaka • M. Spiske
Editors

Tsunamis and Earthquakes in Coastal Environments

Significance and Restoration

Springer

Editors
V. Santiago-Fandiño
Villaviciosa, Asturias, Spain

M. Spiske
Department of Geology and Paleontology
Westfälische Wilhelms-University
Münster, Germany

H. Tanaka
Department of Civil Engineering
Tohoku University
Sendai, Japan

ISSN 2211-0577 ISSN 2211-0585 (electronic)
Coastal Research Library
ISBN 978-3-319-28526-9 ISBN 978-3-319-28528-3 (eBook)
DOI 10.1007/978-3-319-28528-3

Library of Congress Control Number: 2016937980

Printed on acid-free paper

This Springer imprint is published by Springer Nature
The registered company is Springer International Publishing AG Switzerland

Foreword

Coastal zones are valuable areas where sea, land, and atmosphere meet. They are characterized by coastal topography and nearshore waves which drive various dynamic processes to interact with each other. The coastal topography serves as the basis of the environment, in which waves will break and produce turbulent mixing, dissipating wave energy and providing a wealthy coastal zone utilized by many ecosystems for various activities. In the coastal zone, vigorous sediment movement due to nearshore waves continuously changes the topography which protects land from flooding. Coastal environments based on coastal topography are therefore important in coastal hazard mitigation as well as sustainability of society. However, coastal environments are highly variable in broad timescales from minutes to decades, affected by natural and anthropogenic impacts from both the ocean and the land.

Coastal erosion is accelerating on many coasts all over the world. Water pollution and eutrophication are degrading coastal environments in semi-enclosed bays backed by megacities. Understanding the cumulative impact of these types of environmental degradation is difficult since the transport of sediment and nutrients is affected not only by waves, currents, and topography in the coastal zone but also by various inland natural and anthropogenic changes that lead to the increase and decrease of materials delivered to the coast. The impact is sometimes rapid urbanization far from the coastal zone. The response is sometimes delayed as long as several decades.

Large tsunamis and earthquakes are infrequent natural events capable of causing large-scale destruction along coastal areas by heavily altering their physical and environmental characteristics. Although coastlines are naturally highly dynamic systems, these events cause dramatic changes in estuaries, coastal lagoons, tidal flats, wetlands, and beaches resulting in large alterations in their morphology, sediment, depth, water quality, surface area, and flow, as well as inhibiting flora and fauna in a very short period of time. On occasion, the impact is so large that these coastal features even disappear completely.

History abounds with examples of devastating tsunamis that originated from different causes, most commonly from earthquakes. The first record of a Holocene tsunami is related to the Storegga slide, a large submarine landslide in the North Sea around 6000 BC, which heavily impacted the coastline of Norway as sediment was found up to 20 m above sea level. In modern-day history, there are more than 50 records of large tsunamis and related earthquakes, including the most recent events in the Indian and Pacific Oceans (2004 Sumatra, 2010 Chile, and 2011 Japan).

In terms of human life, NOAA states that since 1850, about 420,000 lives have been lost due to tsunamis as local coastal communities and large villages, towns, and even cities have been impacted. Likewise, losses in infrastructure, the economy, and ecosystem services have amounted to trillions of dollars.

Tsunamis like the 2004 Indian Ocean Tsunami and the 2011 Tohoku Tsunami caused catastrophic damage to coastal areas. Many reconstruction processes are being introduced in the affected areas including structure-based countermeasures and nonstructure-based relocation and evacuation plans. In consideration of accelerating reconstruction processes, we need to pay deliberate attention to long-term treatment of coastal environments.

Restoration and reconstruction of damaged areas are the cause of large debates; on one side, arguments for the paramount importance of human protection at all costs, without proper consideration of the environment, have prevailed, while on the other, there are increased calls for a more balanced approach to harmonize measures of protection with the environment in order to allow coastal ecosystems to thrive. The restoration process and reconstruction measures in certain countries could be taken as an example of both, but unfortunately aesthetic and visual restoration is often mistaken for full environmental restoration.

This book deals with impacts of tsunamis and earthquakes on coastal environments. In addition to direct impact and response due to flooding and abrasion, the text covers physical, chemical, and biological responses in coastal morphology, water quality, and ecosystems. Comprehensive descriptions of multi-scale impacts of tsunami and earthquake events, both spatially and temporally, will help us to understand the complicated interactions developed in coastal zones and to achieve the sustainable resilient environment and society with smart post-event recovery. I believe this book will be beneficial to researchers and students in science and engineering as well as policy-makers, urban planning engineers, and coastal managers.

The University of Tokyo
Tokyo, Japan

Shinji Sato

Acknowledgments

The editors would like to express deep appreciation to the authors of the chapters of this book for their most valuable contributions. Their collaboration is an example of of dedication and motivation toward the development of science and knowledge as well as sharing of information in the field of tsunami and earthquakes in coastal areas. Likewise, gratitude is also due to the reviewers for their most valuable comments and suggestions and, last but not least, to all others who in one form or another have contributed to the production of this book.

Contents

Contributors

Iftikhar Ahmed Abbasi Department of Earth Science, College of Science, Sultan Qaboos University, Muscat, Sultanate of Oman

Mohammad Bagus Adityawan Water Resources Engineering Research Group, Institut Teknologi Bandung, Bandung, Indonesia

Department of Civil Engineering, Tohoku University, Sendai, Japan

Bruno Adriano Graduate School of Engineering, Tohoku University, Aramaki, Aoba-ku, Sendai, Japan

C. Andrade IDL and Departamento de Geologia, Faculdade de Ciências da Universidade de Lisboa, Lisboa, Portugal

Ram Avtar Institute for the Advance Study of Sustainability (UNU-IAS), United Nations University, Tokyo, Japan

P.J.M. Costa IDL and Departamento de Geologia, Faculdade de Ciências da Universidade de Lisboa, Lisboa, Portugal

N.P. Dissanayaka Institute of Oceanography, University Malaysia Terengganu (UMT), Kuala Terengganu, Malaysia

Jie Dou Graduate School of Frontier Science, The University of Tokyo, Kashiwa, Japan

Miguel Estrada Centro Peruano Japonés de Investigaciones Sísmicas y Mitigación de Desastres (CISMID), Universidad Nacional de Ingeniería, Lima, Peru

M.C. Freitas IDL and Departamento de Geologia, Faculdade de Ciências da Universidade de Lisboa, Lisboa, Portugal

Yushiro Fujii International Institute of Seismology and Earthquake Engineering (IISEE), Building Research Institute (BRI), Tsukuba, Ibaraki, Japan

R. González-Villanueva Dpto. Xeociencias Mariñas e O.T. (XM-1), Facultade de Ciencias do Mar Universidade de Vigo, Campus As Lagoas Marcosende, Vigo, Pontevedra, Spain

Kenji Harada Center for Integrated Research and Education of Natural Hazards, Shizuoka University, Shizuoka, Japan

Vo Cong Hoang Department of Civil Engineering, Tohoku University, Sendai, Japan

Thuyloi University - Southern Campus, Ho Chi Minh, Vietnam

Kinuko Ito Graduate School of Agricultural Science, Tohoku University, Aobaku, Sendai, Japan

L.P. Jayatissa Department of Botany, University of Ruhuna, Matara, Sri Lanka

Din Mohammed Kakar Department of Geology, Baluchistan University, Quetta, Baluchistan, Pakistan

Ayu Katayama Graduate School of Agricultural Science, Tohoku University, Aobaku, Sendai, Japan

Mohammed Asif Khan Karakorum International University, Gilgit, Pakistan

Naoko Kimura Department of Sustainable Rural Development, Graduate School of Global Environmental Studies, Kyoto University, Kyoto, Japan

K.A.S. Kodikara Department of Botany, University of Ruhuna, Matara, Sri Lanka

Widjo Kongko Coastal Dynamics Research Center, BPPT, Jakarta, Indonesia

Shunichi Koshimura International Research Institute of Disaster Science (IRIDeS), Tohoku University, Aramaki, Aoba-ku, Sendai, Japan

Akira Mano International Research Institute of Disaster Science, Tohoku University, Sendai, Japan

Erick Mas International Research Institute of Disaster Science (IRIDeS), Tohoku University, Aramaki, Aoba-ku, Sendai, Japan

Antonio Mascarenhas CSIR-National Institute of Oceanography, Goa, India

Hideo Matsutomi Department of Civil Engineering, Akita University, Akita, Japan

Yuta Mitobe Department of Civil Engineering, Tohoku University, Sendai, Japan

Norihiro Monna Graduate School of Agricultural Science, Tohoku University, Aobaku, Sendai, Japan

Abdul Muhari Directorate for Coastal and Ocean, Ministry of Marine Affairs and Fisheries, Republic of Indonesia, Jakarta, Indonesia

A.C. Narayana Centre for Earth & Space Sciences, University of Hyderabad, Hyderabad, India

Osamu Nishimura Department of Civil and Environmental Engineering, Tohoku University, Sendai, Japan

M.A. Oliveira IDL and Departamento de Geologia, Faculdade de Ciências da Universidade de Lisboa, Lisboa, Portugal

Min Roh Department of Civil Engineering, Tohoku University, Sendai, Japan

Eko Rudianto Directorate for Coastal and Ocean, Ministry of Marine Affairs and Fisheries, Republic of Indonesia, Jakarta, Indonesia

Enggar Sadtopo Directorate for Coastal and Ocean, Ministry of Marine Affairs and Fisheries, Republic of Indonesia, Jakarta, Indonesia

Takashi Sakamaki International Research Institute of Disaster Science, Tohoku University, Sendai, Japan

Department of Civil and Environmental Engineering, Tohoku University, Sendai, Japan

Youhei Sakurai Department of Civil and Environmental Engineering, Tohoku University, Sendai, Japan

Ahmed Sana Department of Civil Engineering, College of Engineering, Sultan Qaboos University, Muscat, Sultanate of Oman

Vicente Santiago-Fandiño Environmental Advisor, Villaviciosa, Asturias, Spain

B. Satyanarayana Institute of Oceanography, University Malaysia Terengganu (UMT), Kuala Terengganu, Malaysia

Jaya Kumar Seelam CSIR-National Institute of Oceanography, Goa, India

Kazunori Shizuka Graduate School of Agricultural Science, Tohoku University, Aobaku, Sendai, Japan

Hendra Yusran Siry Directorate for Coastal and Ocean, Ministry of Marine Affairs and Fisheries, Republic of Indonesia, Jakarta, Indonesia

Yuriko Takeda International Research Institute of Disaster Science, Tohoku University, Sendai, Japan

Hitoshi Tanaka Department of Civil and Environmental Engineering, Tohoku University, Sendai, Japan

Kaoru Tojo Department of Civil and Environmental Engineering, Tohoku University, Sendai, Japan

Keiko Udo International Research Institute of Disaster Science, Tohoku University, Sendai, Japan

Hideaki Yanagisawa Department of Regional Management, Faculty of Liberal Arts, Tohoku Gakuin University, Izumi-ku, Sendai, Miyagi, Japan

Ali P. Yunus Graduate School of Frontier Science, The University of Tokyo, Kashiwa, Japan

Chapter 1
Revisiting the 2001 Peruvian Earthquake and Tsunami Impact Along Camana Beach and the Coastline Using Numerical Modeling and Satellite Imaging

Bruno Adriano, Erick Mas, Shunichi Koshimura, Yushiro Fujii, Hideaki Yanagisawa, and Miguel Estrada

Abstract On June 23, 2001, a moment magnitude Mw 8.4 earthquake occurred off the southern coast of Peru causing substantial damage to urban and agricultural areas. The tsunami generated by this earthquake reached up to 7 m run-up height and extended over 1.3 km inundation. This paper aims to revisit the impact of the 2001 Peruvian tsunami on the coastal area and its morphology along Camana city. The tsunami source is reconstructed through inversion of tsunami waveform records observed at several tide gauge stations and the impact is analyzed using the numerical result and moderate-resolution satellite images to calculate the inundation features in the coast. Finally we propose the tsunami source model suitable for further analysis of this event through tsunami numerical simulations. In addition,

B. Adriano (✉)
Graduate School of Engineering, Tohoku University,
Aoba 468-1-E301, Aramaki, Aoba-ku, Sendai 980-0845, Japan
e-mail: adriano@geoinfo.civil.tohoku.ac.jp

E. Mas • S. Koshimura
International Research Institute of Disaster Science (IRIDeS), Tohoku University,
Aoba 468-1-E302, Aramaki, Aoba-ku, Sendai 980-0845, Japan
e-mail: mas@irides.tohoku.ac.jp; koshimura@irides.tohoku.ac.jp

Y. Fujii
International Institute of Seismology and Earthquake Engineering (IISEE), Building Research Institute (BRI), 1 Tachihara, Tsukuba, Ibaraki 305-0802, Japan
e-mail: fujii@kenken.go.jp

H. Yanagisawa
Department of Regional Management, Faculty of Liberal Arts, Tohoku Gakuin University,
2-1-1 Tenjinzawa, Izumi-ku, Sendai, Miyagi 981-3193, Japan
e-mail: h-yanagi@izcc.tohoku-gakuin.ac.jp

M. Estrada
Centro Peruano Japonés de Investigaciones Sísmicas y Mitigación de Desastres (CISMID),
Universidad Nacional de Ingeniería, Av. Túpac Amaru No. 1150 – Lima 25 Apartado
Postal 31-250 Lima 31, Lima, Peru
e-mail: estrada@uni.edu.pe

V. Santiago-Fandiño et al. (eds.), *Tsunamis and Earthquakes in Coastal Environments*, Coastal Research Library 14, DOI 10.1007/978-3-319-28528-3_1

environmental changes, in particular the impact to vegetation areas, were evaluated using satellite imagery. An important reduction of agricultural areas due to tsunami impact and soil salinization was confirmed.

Keywords Peru earthquake • Tsunami source • Inversion • Modeling • Remote sensing

1.1 Introduction

The tsunami generated by the 23 June 2001 Peru Earthquake (M_w 8.4) generated a destructive tsunami that struck the southern region of the Peruvian coast. The post-tsunami survey team (ITST: International Tsunami Survey Team) reported tsunami heights, distance of inundation and structural damage (ITST 2001a, b, c). ITST reported that the destruction due to the tsunami was mainly concentrated in Camana along 30 km of a sandy populated beach. The average value of run-up height was 5 m with a maximum value of 7 m. The inundation distance reached between 700 m and 1.3 km inland. As a result, the agricultural fields in the Camana estuary were covered with up to 40 cm of sand deposits (Okal et al. 2002). Official records estimated that at least 57 people were killed by the earthquake and 26 by the tsunami, while 2812 were injured and approximately 60,000 houses were damaged or destroyed (MINSA/OPS 2005). An engineering analysis of the recorded ground motion conducted by Rodriguez-Marek et al. (2010) shows that the earthquake presented peak ground accelerations between 0.04 and 0.34 g for distances from the fault between 70 and 220 km. Within the framework of the project Enhancement of Earthquake and Tsunami Disaster Mitigation Technology in Peru (JST-JICA SATREPS) (Yamazaki et al. 2013), Peruvian and Japanese research teams had conducted field surveys in the affected areas of Camana which remain unreconstructed since 2001. They had found tsunami traces, such as watermarks on walls, still present in remaining destroyed buildings of the urban area to the south of Camana beach (Shoji et al. 2014; Yanagisawa et al. 2011). Furthermore, Spiske et al. (2013) analyzed the changes of tsunami deposits after 7 years from the main event, and found that approximately 25 % of the original sediment layer had been reduced during this period.

The seismological discussion of the 2001 Peru earthquake, including the tsunami impact, had been addressed in detail on several scientific publications (e.g., Audemard et al. 2005; Bilek 2002; Giovanni 2002; Okal et al. 2002; Tavera et al. 2002, 2006). However, few studies had focused on reproducing the tsunami inundation (Adriano et al. 2011; Jimenez et al. 2011). Therefore, in this study, the main purpose is to reproduce, through numerical modeling and satellite image analysis, the tsunami impact of this event to the Camana beach coastline and its morphology. The following discussion is divided in three sections. First, a seismic source model estimated from tsunami data is introduced; this source is suitable for modeling tsunami inundation. The proposed source model is based on tsunami waveform inversion of recorded signals at several tide gauge stations. In the second section, using the calculated seismic source, a non-lineal tsunami numerical modeling is conducted to reproduce the inundation area, the maximum inundation depth, and the tsunami run-up. Finally, in the third section, the tsunami impact to Camana beach coastline is evaluated using the tsunami numerical results and pre- and post-event satellite images.

1.1.1 Developing the Tsunami Source Model

1.1.1.1 Seismic Source Models

The 2001 Peru earthquake occurred on June 23 (15:33:14 local time) along the Peru-Chile trench. The main shock took place at the subduction zone between Nazca and South American plates. The Nazca plate is sub-ducting underneath of the South American plate at a rate of 4.3–4.5 cm/year (Fig. 1.1). This inter-plate contact zone is the longest and one of the most active plate boundaries worldwide (Audemard

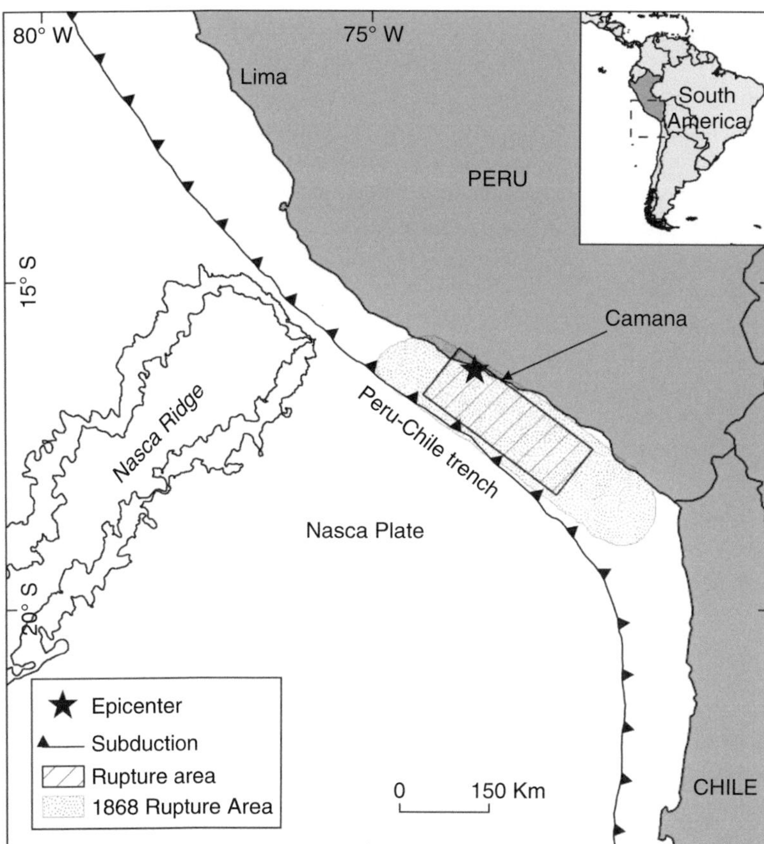

Fig. 1.1 Tectonic setting of southern Peru and northern Chile. The epicenter and rupture area of the 2001 event are shown by the *black star* and the hatched polygon, respectively

et al. 2005) and it has generated several large and catastrophic earthquakes in the past (Dorbath et al. 1990).

The epicenter of the 2001 event was located at 73.64°W 16.26°S (USGS) while the hypocenter was estimated at 73.94 ± 2.5 km west and 16.46 ± 3.9 km south with 30.4 ± 9.5 km depth according to the Geophysical Institute of Peru (IGP) (Tavera et al. 2002; Ocola 2008). Based on the Modified Mercalli Intensity scale the maximum intensity of ground shaking was VII (Tavera et al. 2006). The rupture process duration was estimated by the Global CMT catalog as 86 s and by Kikuchi and Yamanaka (2001) as 107 s. The estimated magnitude varied from M_w 8.2 to M_w 8.4. For instance, Kikuchi and Yamanaka (2001) analyzed teleseismic broadband *P* waves retrieved from 24 seismic stations to determine the general source parameters. They estimated a moment magnitude of M_w 8.2 and a seismic moment of 2.2×10^{21} Nm in a rupture area of 150 km × 240 km. Similarly, Tavera et al. (2002) determined the fault plane orientation and magnitude of Mw 8.2 by analyzing the polarity of *P* waves and body waveform inversion from 15 broadband stations at teleseismic distances. Another analysis of *P* waveforms from 14-recorded seismograms estimated a total seismic moment of 2.4×10^{21} Nm, which gives a moment magnitude of M_w 8.2 (Giovanni 2002). Conversely, the Global CMT catalog reported a seismic moment of 4.7×10^{21} Nm with a moment magnitude of M_w 8.4. This event was followed by intensive aftershock sequences of approximately 400 earthquakes during the first month after the earthquake. The strongest aftershock was on July 7, 2001 with a moment magnitude M_w 7.5 (Bilek 2002; Tavera et al. 2006).

Most of the earthquake source models that have been proposed for the 2001 Peru earthquake are based on the observation and the analysis of seismic data. Conversely, this study contributes on presenting a source model calculated from tsunami waveform inversion as a follow-up estimation from a previous source proposed by Adriano et al. (2012) but using new bathymetry data available and different tide gauges in the area for better approximation.

1.1.1.2 Tsunami Data

The 23 June 2001 Peru tsunami was recorded in several tide gauge stations around the Pacific Ocean (Goring 2002). In this study the tide gauge data were taken from the National Oceanic and Atmospheric Administration (NOAA)/Pacific Marine Environmental Laboratory (PMEL), Center for Tsunami Research, which were provided by the National Oceanic Services (NOS)/NOAA, Field Operation Division, Pacific Regional Office. These data were originally sampled in 1-min and 15-s intervals. According to NOS/NOAA, the data had a minimal editing/reformatting and no quality controls were performed.

For this event, we used tsunami waves recorded at eight tide gauge stations located in Chile and one station located in Peru (Fig. 1.2a). We retrieved the tsunami signal by approximating the tidal wave as a polynomial function and remove it from the original record. Then, the initial time of the earthquake (T_0 = 2001/06/23

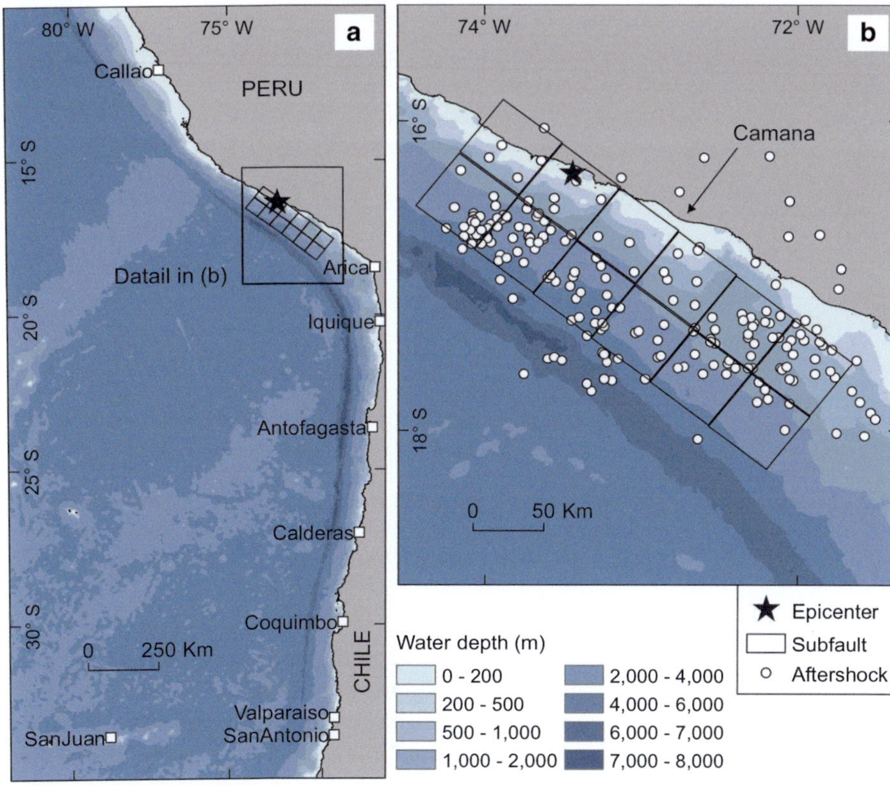

Fig. 1.2 (**a**) Location of the tide gauge stations used for the tsunami waveform inversion (*solid white square*). (**b**) Inferred fault geometry of the 2001 Peru Earthquake from aftershocks distribution

20:33:14 UTC, according to USGS) was subtracted. Finally, the tsunami waveforms were resampled in 1-min interval for tsunami inversion, as shown by red dashed lines in Fig. 1.3.

1.1.1.3 Tsunami Waveform Inversion

The focal parameters of the earthquake source are based on the Global CMT catalog (strike 308°, dip 18°, slip angle 63°). Figure 1.2b shows the epicenter and the extent of aftershocks over a 7-day window following the main-shock. Based on the aftershock distribution a 300 km in length and 100 km width area was set as the rupture zone oriented southeast from the epicenter. In addition, the fault area was divided in 12 subfaults (50 km × 50 km) to cover the aftershock area as shown in Fig. 1.2b. The top depths of subfaults are 14.15 km and 29.60 km southwest northeast from the trench to the coast.

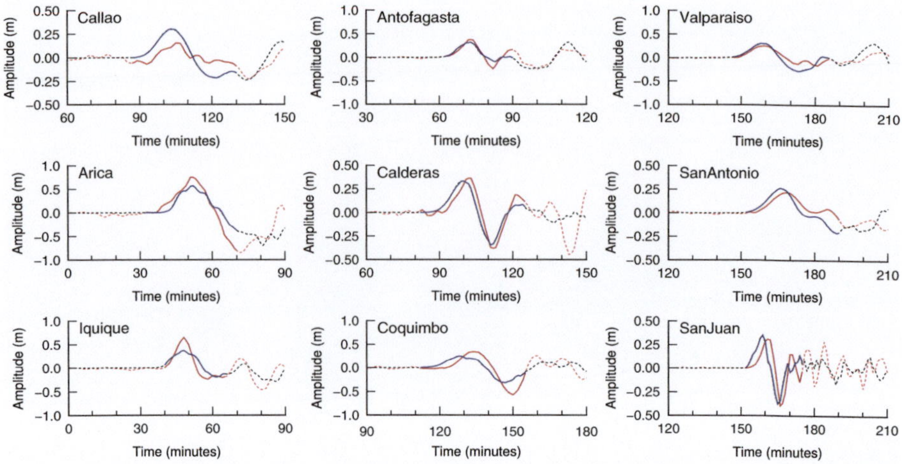

Fig. 1.3 Comparison of the recorded (*red*) and synthetic (*blue*) tsunami waveforms computed from the estimated slip distribution. The time intervals shown in solid lines were used for the inversion; *dashed lines* are shown only for comparison but those were not use in the source model estimation

To calculate the synthetic tsunami waveform at the tide gauge stations, tsunami propagation from each subfault to the stations was calculated using the linear shallow-water approximation (Satake 1995). The computational domain is shown in Fig. 1.2a. The bathymetry data was constructed from the General Bathymetry Chart of the Ocean (GEBCO) 30 arc-seconds grid data. As the initial condition, static deformation of the seafloor is calculated for a rectangular fault model (Okada 1985). In addition, the effect of coseismic horizontal displacement was also included (Tanioka and Satake 1996). The non-negative least square method was used to estimate the slip distribution. The details of the inversion method are described in Fujii and Satake (2007). The calculation of later phases or reflected waves in the tsunami signal is particularly difficult and inaccurate with the coarse resolution in the bathymetry data; therefore, only the first cycle of tsunami waveform was used for the inversion process.

The comparison of the recorded (red curves) and synthetic (blue curves) tsunami waveforms computed from the estimated slip distribution is shown in Fig. 1.3. In general, the synthetic waveforms agree with the observed data at stations located far from the source. Conversely, there is a 15 cm difference on the estimated maximum amplitude at Callao station. Although the maximum amplitude of Callao, Arica, and Iquique stations is not well reproduced, the phases are well estimated. The inversion results are shown in Table 1.1 and Fig. 1.4a. The total seismic moment was calculated from the slip distribution as 4.07×10^{21} Nm (M_w 8.3) using 4.0×10^{10} N/m^2 as average rigidity value in an elastic medium. The inversion result indicates that the maximum slip corresponds to the deepest subfaults (No. 9–10), near Camana beach. The land-level changes calculated from the slip distribution suggest a coastal subsidence of approximately 1.07 m near Camana beach (Fig. 1.4b) while GPS measurements estimated 0.84 cm of land subsidence at Camana (Ocola 2008).

Table 1.1 Earthquake source model obtained from the tsunami waveform inversion

No.	Strike angle (°)	Dip angle (°)	Rake angle (°)	Length (km)	Width (km)	Slip (m)	Depth (km)	Lat. (°)	Lon. (°)
1	308	18	63	50	50	0.05	14.15	−18.25°	−72.20
2	308	18	63	50	50	0.00	14.15	−17.97	−72.58
3	308	18	63	50	50	2.34	14.15	−17.68	−72.95
4	308	18	63	50	50	0.00	14.15	−17.40	−73.32
5	308	18	63	50	50	0.00	14.15	−17.12	−73.69
6	308	18	63	50	50	0.00	14.15	−16.83	−74.07
7	308	18	63	50	50	0.00	29.60	−17.90	−71.92
8	308	18	63	50	50	1.73	29.60	−17.62	−72.30
9	308	18	63	50	50	10.18	29.60	−17.33	−72.67
10	308	18	63	50	50	10.94	29.60	−17.05	−73.04
11	308	18	63	50	50	7.32	29.60	−16.76	−73.41
12	308	18	63	50	50	0.00	29.60	−16.48	−73.79

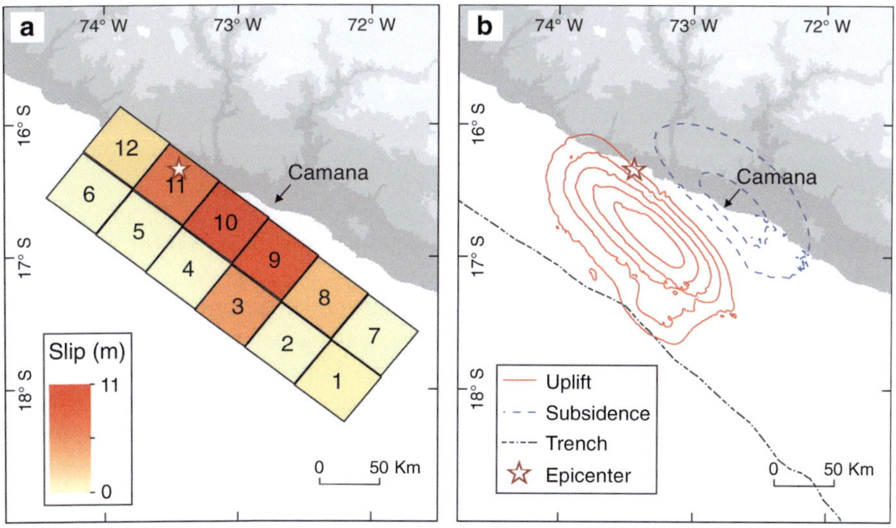

Fig. 1.4 (**a**) Slip distributions estimated by tsunami waveform inversion. The color scale shows the slip value for each subfault. The star shows the epicenter. (**b**) Seafloor deformation computed from the estimated slip distribution. The *red solid contours* indicate uplift with a contour interval of 0.5 m, whereas the *blue dashed contours* indicate subsidence, with a contour interval of 0.5 m

1.1.2 Numerical Simulation of Tsunami Inundation

1.1.2.1 Numerical Model Set-Up

The Tohoku University's Numerical Analysis Model for Investigation of Near-Field Tsunami No.2 (TUNAMI-N2) model, which is based on the non-linear shallow water approximation, was used to conduct the tsunami numerical simulation. A set of non-linear shallow water equations (1.1, 1.2 and 1.33) are discretized by the Staggered Leap-frog finite difference scheme (Imamura 1996). The bottom friction coefficients in the Manning's equation are set according to the land use (Table 1.1).

$$\frac{\partial \eta}{\partial t}+\frac{\partial M}{\partial x}+\frac{\partial N}{\partial Y}=0 \tag{1.1}$$

$$\frac{\partial M}{\partial t}+\frac{\partial}{\partial x}\left(\frac{M^2}{D}\right)+\frac{\partial}{\partial y}\left(\frac{MN}{D}\right)=-gD\frac{\partial \eta}{\partial x}-\frac{gn^2}{D^{7/3}}\mathrm{M}\sqrt{M^2+N^2} \tag{1.2}$$

$$\frac{\partial N}{\partial t}+\frac{\partial}{\partial x}\left(\frac{MN}{D}\right)+\frac{\partial}{\partial y}\left(\frac{N^2}{D}\right)=-gD\frac{\partial \eta}{\partial y}-\frac{gn^2}{D^{7/3}}\mathrm{N}\sqrt{M^2+N^2} \tag{1.3}$$

where

$$M = \int_{\eta}^{-h} udz \tag{1.4}$$

$$N = \int_{\eta}^{-h} vdz \tag{1.5}$$

$$D = \eta + h \tag{1.6}$$

M and *N* are the discharge flux of *x* and *y* direction, respectively, η is the water level and *h* is the water depth above the mean sea level.

The computational domain is divided into four subdomains to construct a nested grid system, as shown in Fig. 1.5. The grid size, which extends from the earthquake source region to the coast of Camana coast, varies from 30 to 810 m. The bathymetry data for the first to the third domains were interpolated from the GEBCO 30 arc-seconds grid data. The merged topography and bathymetry grids for the fourth domain were constructed from the nautical chart provided by the DHN, Navy of Peru (Dirección de Hidrografía y Navegación in Spanish) and the land elevation data obtained from the Thermal Emission and Reflection Radiometer (ASTER) sensor with 1 arc-second grid resolution.

For the tsunami inundation model, the land resistance during tsunami penetration is considered by setting a specific roughness coefficient to the land cover (Table 1.2) (Aburaya and Imamura 2002; Dutta et al. 2007; Kotani et al. 1998). Figure 1.6 shows the spatial distribution of Manning's roughness coefficient (*n*) in the computational domain of tsunami inundation model. The *n*-distribution was obtained through the unsupervised classification of moderate resolution satellite image. In addition, the coastline of Camana beach and surroundings were estimated from the analysis of satellite images. The detail of this step is discussed in Sect. 1.4.1.

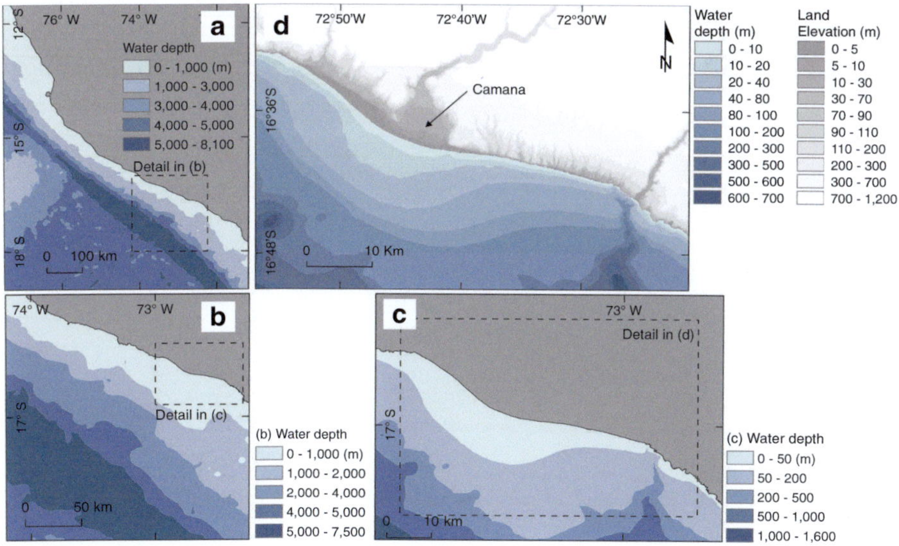

Fig. 1.5 The computational domain for the model of tsunami propagation and inundation to the coastal area of Camana city. The grid size varies from 810, 270, 90 to 30 m, as shown in the nested grid system detailed by (**a–d**)

Table 1.2 Manning's roughness coefficient (n) values according to (Kotani et al. 1998)

Smooth ground	0.020
Shallow water area or natural beach	0.025
Vegetated area	0.030
Populated area	0.040

1.1.2.2 Results and Validation of Tsunami Inundation Model in Camana Coast

As shown in Fig. 1.5, the topography in Camana coast is predominantly of low lands penetrating as much as 1.0 km. Following the plain beach area, steep hills can be observed in the north and south from the river mouth (Fig. 1.5d). This topography gives a natural barrier against the tsunami penetration. According to the satellite imagery analysis, almost 5 km inland from the coastline of Camana beach is used for agriculture. In addition, based on the model results, the tsunami penetrated 1.0–1.5 km inland in this area.

Results from simulation were validated through the comparison with field surveyed inundation data. Figure 1.7 shows the spatial distribution of modeled maximum tsunami inundation depths. The maximum inundation depth in the computational domain is 8.5 m. The tsunami penetration at Playa Pucchun and south of Camana beach (La Punta – Las Cuevas) agrees with the observed data (ITST 2001a, b, c; Dengler 2001). It was observed that the urban area located to the south of Camana beach, between the La Punta and the Las Cuevas, was inundated

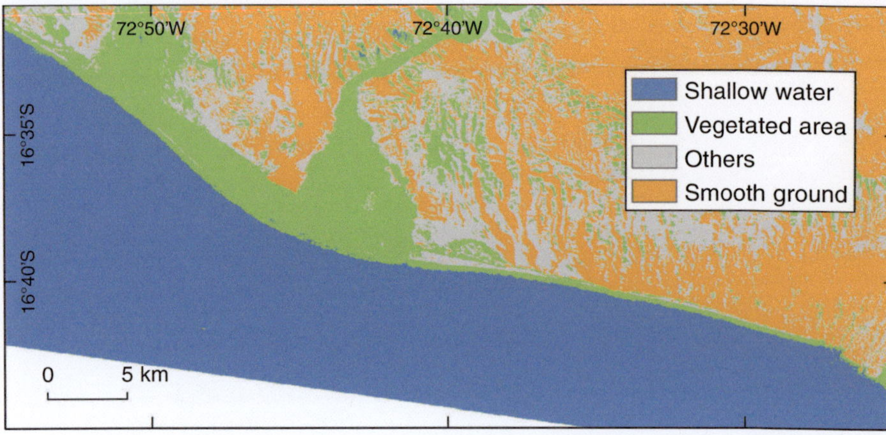

Fig. 1.6 Spatial distribution of roughness coefficient, according to the land-use condition, inferred from the unsupervised classification of pre-tsunami satellite imagery (Landsat-5 TM)

with 3–4 m (Fig. 1.7d). On the other hand, the agricultural areas to the north of Camana beach were inundated with 2–3 m (Fig. 1.7b, c). Except for areas in the north of Camana beach (Playa la Chira), where the calculated tsunami penetration slightly overestimates the limits observed by the ITST 2001 survey, simulation results are consistent with in-situ measurement data. Discrepancies in the north part of Camana are due to the limitation of the shallow water approximation added to the lack of local bathymetry information within the study area, in particular near the Playa La Chira.

The modeled tsunami inundation is validated by K and κ coefficients proposed by Aida (1978) According to the ITST 2001 survey, 20 points of local inundation depth were measured around Camana beach (Fig. 1.7a). Figure 1.8 shows a lineal comparison of simulated and measured tsunami inundation depth. Based on the suggested guidelines provided by JSCE (2002) ($0.95<K<1.05$ and $\kappa<1.45$), our results can be evaluated as suitable ($K=1.02$ and $\kappa<1.07$) to be used for evaluating tsunami impact along Camana beach coastline and its morphology.

1.1.3 Impact of the Tsunami Inundation in Camana Coast

1.1.3.1 Method

Satellite images were selected based on its acquisition time and free-cloud coverage in the study area. Unfortunately there was no available image acquired within a short time after or before the tsunami event. Nevertheless, we selected three Landsat-5 Thematic Mapper (TM) images scenes (Path 04, Row 71; according to the World Reference System 2). Two pre-event images taken on June 21, 1999, almost 2 years before the event, and on June 10, 2001, almost 2 week before the

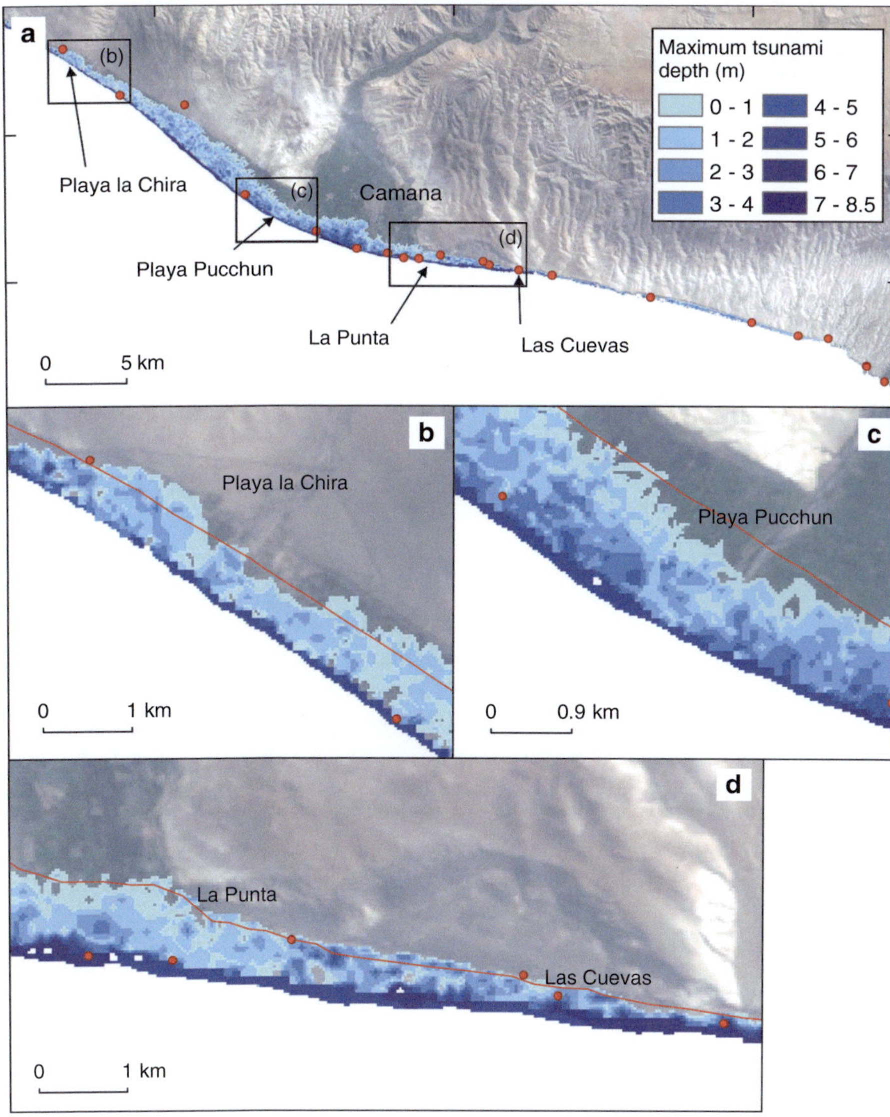

Fig. 1.7 Spatial distribution of modeled inundation depth. The *solid red line* and *red circles* indicate the tsunami inundation limit and measured inundation depth (ITST 2001a; Dengler 2001). (**a**) Distribution of inundation depth in the computational domain. (**b–d**) inundation depth at the most affected areas in Camana according to the ITST 2001. Background image corresponds to the pre-event Landsat TM acquired on June 10, 2001

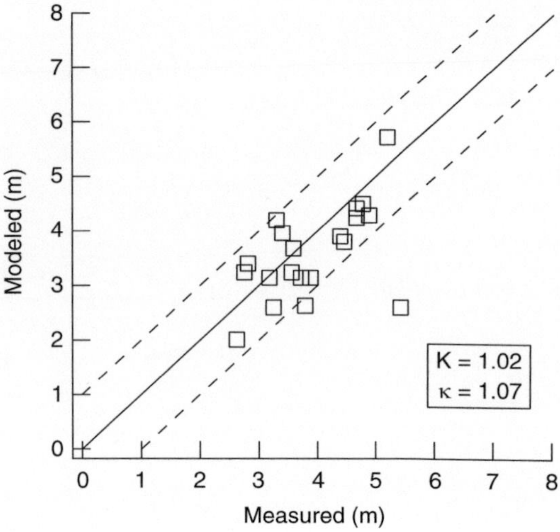

Fig 1.8 Comparison of the model results and the field survey data, in terms of the local inundation depths

tsunami event. The third image was taken on June 21, 2002, almost 1 year after the event. Images were selected based on its acquisition time and month to seek few days of difference with the time of the event and to reduce the seasonal sensitivity of optical satellite images (Zhang et al. 2013).

Landsat-5 TM image data files consist of seven spectral bands, including one thermal infrared (band 6). The resolution is 30 m for bands 1–5, and 7, detailed information about the Landsat product can be found at USGS-Landsat Missions website. Several pre-processing algorithms were carried out to the images in order to extract quantitative information about the features on the land surface. In addition, the images were atmospherically corrected by transforming uncalibrated pixel value (digital number) into surface reflectance images (Chander et al. 2004, 2009; Markham and Helder 2012).

1.1.4 Remote Sensing Analysis

Two analyses were performed to delineate the coastline of Camana beach and surroundings and to investigate the changes of the coastal morphology in terms vegetation. First, using the images taken on June 10, 2001, we estimated the coastline by calculating the normalized difference water index (NDWI) and the normalized difference vegetation index (NDVI) (Jackson et al. 2004). Then, the coastline was manually delineated from the index-images taking as reference the true-color composite of the same image. Second, a visual interpretation of the coast of Camana based on the RGB-color composite images and NDVI-images was conducted. The natural-like color composite image was constructed by assigning the Band 7 to the red channel, Band 5 to the green channel, and Band 3 to the blue channel. In this

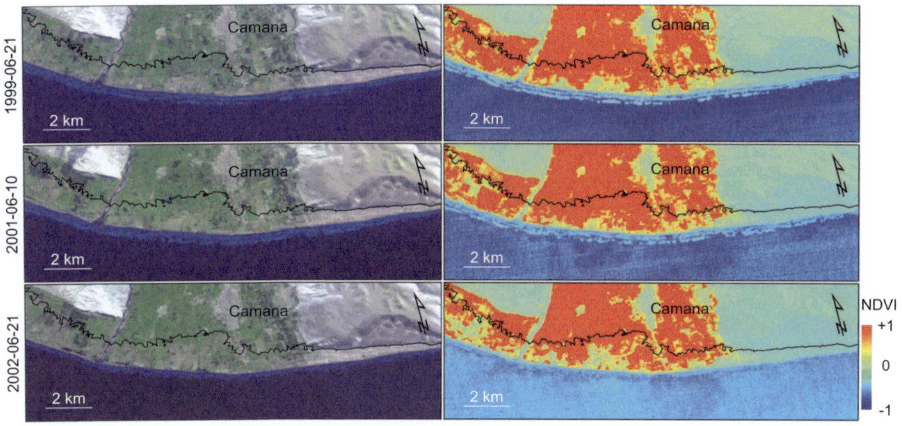

Fig. 1.9 Right panels show the nature-like composite of Landsat-5 TM images. *Left panels* show the NDVI image calculated from Landsat-5 TM dataset (R: band7, G: band5, and B: band3). *The black line* shows the calculated tsunami inundation limits

composition the vegetation appears in dark and light green, urban features are white, gray, cyan or purple. In addition, by absorbing the mid-infrared bands, it provides well-defined coastlines and highlighted sources of water within the image.

Figure 1.9 shows the multi-temporal comparison of the Landsat-5 TM imagery. The extension of the calculate inundation area is shown by the black line. The presence of healthy or unhealthy vegetation can be detected by applying the NDVI index to the temporal satellite images of the areas affected by the tsunami in Camana, as shown in Fig. 1.9. In addition, from optical satellite imagery it is possible to follow the changes on coastline shape and morphology of the area. In the case of Camana coast, from the image taken 1 year after the disaster, approximately 80.1 % of healthy vegetation area had been reduced, while between the two pre-disaster images (1999–2001) the difference in vegetation area in the same season is 2.2 %. Such high reduction of vegetation even 1 year after the disaster can be explained not only by the damage occurred to agricultural areas in June 2001, but also due to the sand deposition by tsunami as reported by Jaffe et al. (2003) which ruined crops with salinization of soil (Olcese 2004), as shown in Fig. 1.10.

1.1.5 Concluding Remarks

Using the recorded tsunami waves at nine tide gauge stations, we estimated the seismic source distribution of the June 23, 2001 Peru Earthquake by using tsunami waveform inversion. The rupture area was constrained based on the aftershock distribution of 7-days. The aftershocks distribution indicated that the 2001 Peru tsunami source was about 300 km in length, extending from the epicenter to the southeast. The maximum slip was approximately 10–11 m corresponding to the

Fig. 1.10 Damage to crops near Playa Pucchun. Similar condition to farmlands along the Camana coast was reported by field survey reports. Salinization and damage to irrigation made difficult to restore vegetation in this areas

deepest subfault, near Camana beach. The total seismic moment was calculated from the slip distribution as 4.07×10^{21} Nm (M_w 8.3).

In addition, based on moderate resolution bathymetry and topography data, tsunami inundation modeling using the estimated tsunami source was presented. The results of simulated local inundation depth and inundation area were consistent with the measured and observed field data. The model verified that the urban area on the south and the agricultural areas on the north of Camana beach were inundated with 3–4 m and 2–3 m, respectively.

Finally, combining the results of inundation area and the analysis of pre- and post-event satellite images, the observed reduction of agricultural areas due to tsunami impact and soil salinization was detected. From the image taken 1 year after the disaster, approximately 80.1 % of healthy vegetation area had been reduced.

References

Aburaya T, Imamura F (2002) The proposal of a tsunami run–up simulation using combined equivalent roughness. Annu J Coast Eng Japan Soc Civ Eng 49:276–280

Adriano B, Koshimura S, Fujii Y (2011) Validation of tsunami inundation modeling for the June 23, 2001 Peru Earthquake. In: 8th international conference on urban earthquake engineering, pp 1800–1804

Adriano B, Koshimura S, Fujii Y (2012) Tsunami source and inundation modeling of the June 2001 Peru earthquake. In: 9th international conference on urban earthquake engineering/4th Asia conference on earthquake engineering, pp 2061–2065. http://ares.tu.chiba-u.jp/peru/pdf/output/cp/9th_CUEE_Bruno_et_al.pdf. Accessed 9 Feb 2015

Aida I (1978) Reliability of a tsunami source model derived from fault parameters. J Phys Earth 26(1):57–73. doi:10.4294/jpe1952.26.57

Audemard MF, Gómez JC, Tavera HJ, Orihuela GN (2005) Soil liquefaction during the Arequipa Mw 8.4, June 23, 2001 earthquake, southern coastal Peru. Eng Geol 78(3-4):237–255. doi:10.1016/j.enggeo.2004.12.007

Bilek SL (2002) Analysis of the 23 June 2001 Mw 8.4 Peru underthrusting earthquake and its aftershocks. Geophys Res Lett 29(20), CitelD1960. doi:10.1029/2002GL015543

Chander G, Helder DL, Markham BL, Dewald JD, Kaita E, Thome KJ, Micijevic E, Ruggles TA (2004) Landsat-5 TM reflective-band absolute radiometric calibration. IEEE Trans Geosci Remote Sens 42:2747–2760. doi:10.1109/TGRS.2004.836388

Chander G, Markham BL, Helder DL (2009) Summary of current radiometric calibration coefficients for Landsat MSS, TM, ETM+, and EO-1 ALI sensors. Remote Sens Environ 113(5):893–903. doi:10.1016/j.rse.2009.01.007

Dengler L (2001) Impacts of the June 23, 2001 Peru tsunami [online]. Available from: http://www.gweaver.net/techhigh/projects/Waves/Tsunami Project/short_peru_report.pdf. Accessed 5 Feb 2015

Dorbath L, Cisternas A, Dorbath C (1990) Assessment of the size of large and great historical earthquakes in Peru. Bull Seismol Soc Am 80(3):551–576. http://www.bssaonline.org/content/80/3/551.abstract. Accessed May 2015

Dutta D, Alam J, Umeda K, Hayashi M, Hironaka S (2007) A two-dimensional hydrodynamic model for flood inundation simulation: a case study in the lower Mekong river basin. Hydrol Process 21(9):1223–1237. doi:10.1002/hyp.6682

Fujii Y, Satake K (2007) Tsunami source of the 2004 Sumatra-Andaman earthquake inferred from tide gauge and satellite data. Bull Seismol Soc Am 97(1A):S192–S207. doi:10.1785/0120050613

Giovanni MK (2002) The June 23, 2001 Peru earthquake and the southern Peru subduction zone. Geophys Res Lett 29(21):2018. doi:10.1029/2002GL015774

Goring DG (2002) Response of New Zealand waters to the Peru tsunami of 23 June 2001. N Z J Mar Freshw Res 36(1):225–232. doi:10.1080/00288330.2002.9517081

Imamura F (1996) Review of tsunami simulation with a finite-difference method. In: Yeh H, Liu P, Synolakis C (eds) Long-wave runup models, World Scientific, pp 25–42. ISBN: 981-02-2909-7

International Tsunami Survey Team (2001a) Impacts of the 2001 Peru tsunami in Camana. In: International tsunami symposium proceedings, p 409

International Tsunami Survey Team (2001b) Report of the June 23, 2001 Peruvian tsunami field survey of the International Tsunami Survey Team (ITST). http://www.usc.edu/dept/tsunamis/peru01/webpages/peru2001tsun.html. Accessed April 2015

International Tsunami Survey Team (2001c) The Peruvian tsunami of 23 June 2001: preliminary report by the International Tsunami Survey Team. In: International tsunami symposium proceedings, pp 377–378

Jackson TJ, Chen D, Cosh M, Li F, Anderson M, Walthall C, Doriaswamy P, Hunt ER (2004) Vegetation water content mapping using Landsat data derived normalized difference water index for corn and soybeans. Remote Sens Environ 92:475–482

Jaffe BE, Gelfenbaum G, Rubin D, Peters R, Anima R, Swensson M, Olcese D, Anticona LB, Gomez JC, Riega PC (2003) Identification and interpretation of tsunami deposits from the June 23, 2001 Peru tsunami. In: International conference on coastal sediments, p 13

Japan Society of Civil Engineers (2002) Tsunami assessment method for nuclear power plants in Japan, 72p

Jimenez C, Adriano B, Koshimura S, Fujii Y (2011) The tsunami of Camana 2001. In: 8th international conference on urban earthquake engineering, pp 1567–1571

Kikuchi M, Yamanaka Y (2001) Near coast of Peru earthquake (Mw 8.2) on June 23, 2001. http://wwweic.eri.u-tokyo.ac.jp/topics/200106232033/index.html. Accessed April 2015

Kotani M, Imamura F, Shuto N (1998) Tsunami run-up simulation and damage estimation by using GIS. In: Proceedings of coastal engineering, JSCE 45, pp 356–360

Markham BL, Helder DL (2012) Forty-year calibrated record of earth-reflected radiance from Landsat: A review. Remote Sens Environ 122:30–40. doi:10.1016/j.rse.2011.06.026

MINSA/OPS Ministerio de Salud – Organizacion Panamericana de la Salud (2005) El Terremoto y Maremoto del Sur del Peru 2001. http://bvpad.indeci.gob.pe/doc/pdf/esp/doc36/doc36_contenido.pdf. Accessed Feb 2015

Ocola L (2008) Southern Perú coseismic subsidence: 23 June 2001 8.4-Mw earthquake. Adv Geosci 14:79–83. doi:10.5194/adgeo-14-79-2008

Okada Y (1985) Surface deformation due to shear and tensile faults in a half-space. Bull Seismol Soc Am 75(4):1135–1154. http://bssa.geoscienceworld.org/content/75/4/1135.short. Accessed Aug 2012

Okal E, Dengler L, Araya S, Borrero J, Gomer B, Koshimura S, Laos G, Olcese D, Ortiz M, Swensson M, Titov V, Vegas F (2002) Field survey of the Camaná, Perú tsunami of 23 June 2001, Seismol. Res Lett 73(6):907–920. doi:10.1785/gssrl.73.6.907

Olcese D (2004) Estudio del origen y consecuencias del tsunami del 23 de Junio del 2001 en la costa sur del Peru, National University of Engineering, Peru

Rodriguez-Marek A, Bay JA, Park K, Montalva GA, Cortez-Flores A, Wartman J, Boroschek R (2010) Engineering analysis of ground motion records from the 2001 M w 8.4 Southern Peru earthquake. Earthq Spectra 26(2):499–524. doi:10.1193/1.3381172

Satake K (1995) Linear and nonlinear computations of the 1992 Nicaragua earthquake tsunami. Pure Appl Geophys PAGEOPH 144(3–4):455–470. doi:10.1007/BF00874378

Shoji G, Shimizu H, Koshimura S, Estrada M, Jimenez C (2014) Evaluation of tsunami wave loads acting on walls of confined-masonry-brick and concrete-block houses. J Disaster Res 9(6):976–983

Spiske M, Piepenbreier J, Benavente C, Bahlburg H (2013) Preservation potential of tsunami deposits on arid siliciclastic coasts. Earth Sci Rev 126:58–73. doi:10.1016/j.earscirev.2013.07.009

Tanioka Y, Satake K (1996) Tsunami generation by horizontal displacement of ocean bottom. Geophys Res Lett 23(8):861. doi:10.1029/96GL00736

Tavera H, Buforn E, Bernal I, Antayhua Y, Vilacapoma L (2002) The Arequipa (Peru) earthquake of June 23, 2001. J Seismol 6(6):279–283. doi:10.1023/A:1015698621075

Tavera H, Fernández E, Bernal I, Antayhua Y, Agüero C, Rodríguez HSS, Vilcapoma L, Zamudio Y, Portugal D, Inza A, Carpio J, Ccallo F, Valdivia I (2006) The southern region of Peru earthquake of June 23rd, 2001. J Seismol 10(2):171–195. doi:10.1007/s10950-006-9014-2

U.S. Geological Survey (2015) Ladsat program, Landsat Missions. http://landsat.usgs.gov. Accessed Feb 2015

Yamazaki F, Zavala C, Nakai S, Koshimura S, Saito T, Midorikawa S, Aguilar Z, Estrada M, Bisbal A (2013) SATREPS project on enhancement of earthquake and tsunami disaster mitigation technology in Peru. J Disaster Res 9(6):916–924

Yanagisawa H, Koshimura S, Yagi Y, Fujii Y, Shoji G, Jimenez C (2011) The tsunami vulnerability assessment in Peru using the index of potential tsunami exposure. In: 8th international conference of urban earthquake engineering, pp 1591–1595

Zhang H, Zhang Y, Lin H (2013) Seasonal effects of impervious surface estimation in subtropical monsoon regions. Int J Digital Earth 7(9):746–760. doi:10.1080/17538947.2013.781241

Chapter 2
Imprints of the AD 1755 Tsunami in Algarve (South Portugal) Lowlands and Post-impact Recovery

P.J.M. Costa, M.A. Oliveira, R. González-Villanueva, C. Andrade, and M.C. Freitas

Abstract The AD 1755 tsunami was the most devastating tsunami that affected Atlantic Europe in historical times. In this work we summarize its sedimentological signatures in lowlands (Martinhal, Barranco, Furnas, Boca do Rio, Salgados-Alcantarilha) of the Algarve coast that contrast in geologic and geomorphological settings and sediment abundance. We found remarkable similarities between tsunami deposits and the materials available for transport at the coast prior to the tsunami. A number of 2–4 m high and 30–70 m wide scarps (i.e. steep slopes in dunes resulting from erosion) are the only erosive geomorphic signature preserved in the study areas (Boca do Rio and Salgados-Alcantarilha). Recovery of the coastal system to pre-event conditions, inferred from the documentary records, is evaluated and analyzed in terms of sediment availability and supply, climate, hydrodynamic regime and geomorphic setting. Sediment-starved pocket beaches bypassed landward most of the sediment previously accumulated in the coastal system, failing to recover the pre-event morphology. Moreover, and solely in Martinhal, the abrupt morphological changes translated in increased and lasting permeability of the barrier and adjacent wetland to storms. The Alcantarilha-Salgados beach-dune system, in moderate sand supply, was extensively scarped but not fully overtopped. Here, recovery was partly achieved by destabilization of the remnant dune, formation and (limited) advance of parabolic dunes following the 1755 event. A schematic conceptual model is presented summarizing: the pre-event conditions; the depositional and geomorphological features directly related with the tsunami impact; and the post-event geomorphological adaptation. In the studied cases post-event recovery seems to be reduced mainly due to a conjugation of pre-event geomorphological setting and low sediment input.

P.J.M. Costa (✉) • M.A. Oliveira • C. Andrade • M.C. Freitas
IDL and Departamento de Geologia, Faculdade de Ciências da Universidade de Lisboa, Edifício C6, Campo Grande 1749-016, Lisboa, Portugal
e-mail: ppcosta@ciencias.ulisboa.pt

R. González-Villanueva
Dpto. Xeociencias Mariñas e O.T. (XM-1), Facultade de Ciencias do Mar Universidade de Vigo, Campus As Lagoas Marcosende, Vigo, 36310, Pontevedra, Spain

V. Santiago-Fandiño et al. (eds.), *Tsunamis and Earthquakes in Coastal Environments*, Coastal Research Library 14, DOI 10.1007/978-3-319-28528-3_2

The need to conduct further studies in the impacts of palaeotsunamis is evidenced here, especially because older events allow a large time window to observe post-tsunami adaptation.

Keywords Geomorphology • Tsunami deposit • Sand barrier • Erosional features • Recovery

2.1 Introduction

The first studies relating the coastal sedimentological record with (pre)historic tsunamis were conducted by Atwater (1987) and Dawson et al. (1988). Since then many papers have been published on tsunami deposits and on related transport and depositional processes (e.g. Dawson and Stewart 2007; Paris et al. 2009; Chagué-Goff et al. 2011; Bahlburg and Spiske 2011). The geomorphological impacts of both present-day tsunamis and palaeotsunamis have also been addressed although less explored (e.g. Andrade 1992; Dawson 1994; Goff 2008; Goff et al. 2009; Atwater et al. 2013, 2014; Kain et al. 2014; Catalán et al. 2014). These works, and particularly Goff et al. (2009), contributed to catalogue geomorphological features produced or triggered by tsunami impacts (e.g. sand/gravel sheets, sand washovers, scour, hummocky topography, dunes, beach/backbeach scarping, etc.).

One key factor to achieve an understanding of (palaeo)tsunami events is the reliability of their sedimentological and geomorphological record. In the respective coastal setting prior to the event this depends on the nature and the amount of sediment available in each coastal cell. The former determines the ability to produce a conspicuous imprint in onshore coastal stratigraphy; the latter is crucial for post-tsunami coastal recovery and preservation of the imprint. The comprehension of the geomorphological and sedimentological signatures of (palaeo)tsunamis is favored by an approach at broad spatial scales in order to incorporate other modulating key factors such as climate variability, hydrodynamic regimes, sediment budgets or onshore geomorphological setting.

Studies on both the coastal geomorphological consequences of the AD 1755 tsunami and post-event response are scarce, despite its relevance to Europe, in particular to SW Iberia. Several studies have focused in the characterization of sand layers deposited by that tsunami in Algarve (South Portugal) lowlands (e.g. Hindson and Andrade 1999; Kortekaas and Dawson 2007; Oliveira et al. 2009; Costa et al. 2012a), in contrast with a single regional study on the geomorphological impacts and post-event recovery in the eastern Algarve (Andrade 1992).

The present study addresses the sedimentological and geomorphological impacts of the AD 1755 tsunami in coastal lowlands of the western and central Algarve, and discusses the post-event recovery efficiency of the coastal systems while at the same time aims at: (a) summarize, at a regional scale, the depositional features laid down by the tsunami; (b) scrutinize the geomorphological impacts caused by the tsunami inundation; (c) analyze the coastal system response in a region strongly impacted by the most devastating tsunami that affected the coasts of Atlantic Europe in historical times.

2.2 Regional and Local Setting

2.2.1 *Regional Geomorphology*

The southern Portuguese coast is geomorphologically highly asymmetric mainly due to pronounced contrasts in the outcropping geology. In its westernmost section, the coastline is strongly irregular, developing in resistant Mesozoic limestone, the geotectonic setting favoring the development of high (~70 m), steep (at times, plunging) cliffs, occasionally interrupted by small pocket beaches. These beaches accumulate sand over gravel and occur in relation with deep and narrow embayments at the outlet of small intermittent streams, which run through deeply incised, tectonically controlled canyons. The streams drain small drainage basins and the sand input to the coast is reduced. Further east, the central Algarve coast develops in Miocene sandy calcareous rocks, softer than the Mesozoic materials, forming low to 40 m-high cliffs. Sand supply is moderate and ensured by streams and cliff erosion. Here, streams and rivers present larger drainage basins and outlet in lagoons, barred estuaries and wide bays with long sandy beaches frequently backed by cliffs and, in cases, by dunes. In contrast, the eastern Algarve coast shows higher abundance in sand and comprises only aggradation forms, such as one large barrier island-lagoon complex and a coastal plain with continuous broad beaches backed by dunes (Andrade 1990) (Fig. 2.1).

Coastal lowlands of the Algarve evolved throughout the Holocene in three stages: inundation of a dissected surface by the Holocene transgression forming drowned morphologies, such as rias; following sea level stabilization close to the present-day level circa 5 ka cal BP, barriers developed, enclosing estuaries and lagoons; during the last c. 3–5 k years marine and terrestrial sediment promoted partial to complete infilling of the back-barrier wetlands.

2.2.2 *Study Areas*

The study sites (Figs. 2.2 and 2.3) were restricted to locations where onshore sedimentary signature of the AD 1755 tsunami has been confirmed. Martinhal lowland (Figs. 2.1 and 2.2a) comprises a triangular shaped high-intertidal alluvial plain/ lagoon at ca. 2 m mean sea level (msl), and a sand barrier with a linear foredune reaching ~12 m (msl). The barrier is interrupted by an ephemeral inlet and is overtopped by storm waves, with washovers extending landward up to 100 m of the dune. Extreme storms in the Algarve coast are related to SW waves and typically associate with wave heights of 2–3 m and 7–8 s period. Once a year, wave heights of 4 m can occur and of 5 m every 5 years (Pires 1985). These events are capable of invading the Martinhal lowland but fail to flood the other studied sites.

Barranco and Furnas lowlands (Fig. 2.2b, c) develop at the mouth of constricted flat floored canyons with steep slopes, comprising supratidal narrow alluvial plains and a barrier. The latter comprises small incipient foredunes and climbing dunes

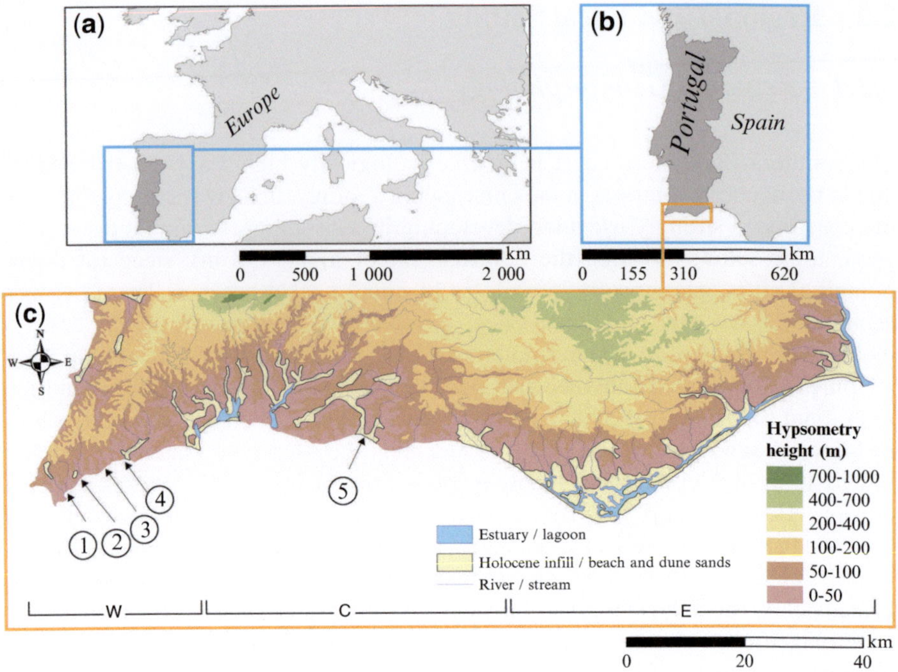

Fig. 2.1 (**a**) Location of Portugal within Europe; (**b**) Location of Algarve in Portugal; (**c**) Algarve hypsometry, with location of rivers, water bodies, Holocene aggradation forms and study areas: *1* Martinhal, *2* Barranco, *3* Furnas, *4* Boca do Rio, *5* Alcantarilha and Salgados

backing small pocket beaches composed of a thin veneer (<1 m) of sand covering shingle. Boca do Rio lowland (Fig. 2.2d) corresponds to a N-S elongated flat-floored valley consisting of an active supratidal flood plain, separated from the sea by a low shingle and sandy barrier that, together with a rock spur, prevent wave overtopping during storms (Hindson and Andrade 1999; Hindson et al. 1999).

The coastal area of Alcantarilha and Salgados lowlands (Fig. 2.3) corresponds to a 6 km long continuous and wide sandy embayed-beach backed by a multiple-ridged dune. The beach/dune barrier leans against Miocene bedrock where it forms an interfluve separating the lowlands. Laterally, the barrier (and dune) is interrupted and limited by two ephemeral inlets. Further inland, muddy fluvial sediments almost fully chocked the Alcantarilha palaeoestuary and actively infill the Salgados lagoonal system.

2.2.2.1 Regional Late Holocene Stratigraphy

Four lithostratigraphic units represent the regional Late Holocene infill of Martinhal, Boca do Rio, Alcantarilha and Salgados lowlands (Figs. 2.4a, b); in Barranco and Furnas the basal unit has not yet been recognized.

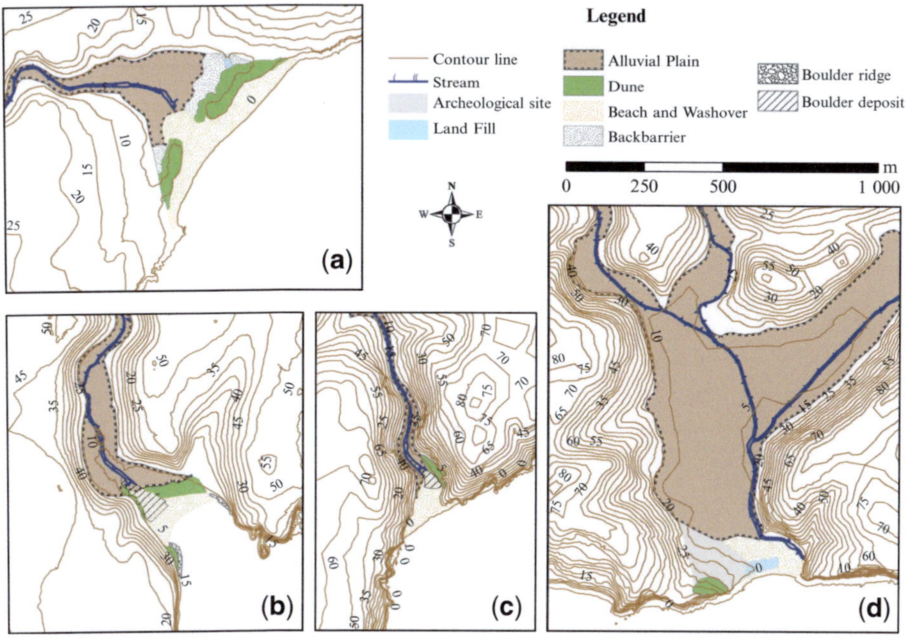

Fig. 2.2 Geomorphological sketch of the studied lowlands: *a* Martinhal, *b* Barranco, *c* Furnas, *d* Boca do Rio

Basal Unit 4 consists of medium sand and gravel with marine shell fragments representing deposition in drowned estuarine environments open to marine influence, predating the barriers (Hindson et al. 1999; Kortekaas and Dawson 2007; Costa et al. 2012a).

Unit 3 consists of alluvial/estuarine/lagoonal muds, with few sand layers towards the base, representing low energy sedimentation in a restricted environment that followed barrier formation.

Unit 2 corresponds to a widespread layer of marine sand with shell fragments, which rises and thins inland. Muddy rip-up clasts are frequent and floating limestone boulders may exist at its base, which is erosional, reflecting the high energy deposit of the AD1755 tsunami inundation. In Barranco and Furnas lowlands unit 2 is <10 cm thick and mainly comprises limestone boulders (0.3–1 m a-axis). These boulders were transported hundreds of meters inland and upwards in relation to the present day shoreline (Costa et al. 2011).

Unit 1 caps the sequence and atop represents the present-day depositional system. It consists of alluvial muds in Boca do Rio, Barranco, Furnas and Alcantarilha, lagoonal muds in Salgados and alternating fluvial mud and marine sand – due to occasional storm-driven sand deposition disturbing the permanent regime of muddy sedimentation – in the case of Martinhal (see Fig. 2.1 for locations).

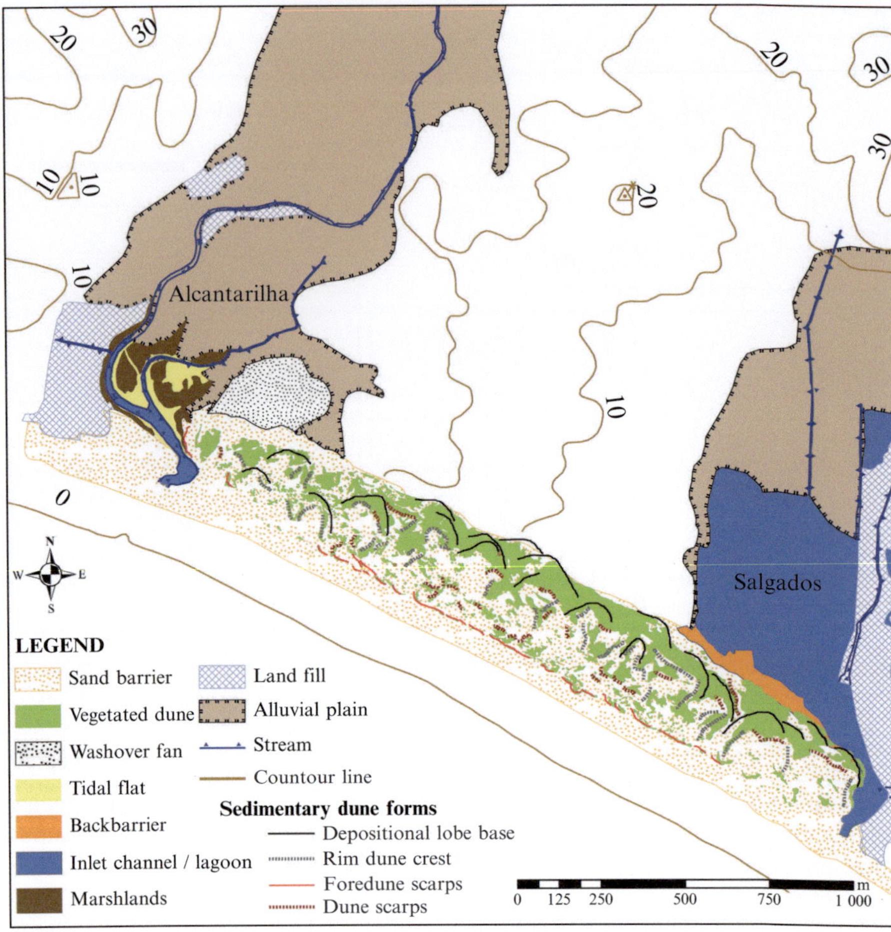

Fig. 2.3 Geomorphological interpretation of Alcantarilha -Salgados lowlands and sandy coastal forms in between. Sedimentary dune forms originated after the tsunami event

2.2.3 *Results and Discussion*

2.2.3.1 Historical and Sedimentological Signature of the AD 1755 Tsunami

The tsunami inundation in Martinhal lowland was described by Lopes (1841) as: "*The sea flooded a beach called Mortinhal, facing eastward, by about ½ league ripping off vineyards and leaving the land as a beach, covered with several types of fish and large boulders of which one, weighting more than 300 arrobas* [~4400 kg] *showed many shellfish stuck on its surface.*"

Andrade et al. (1997) and Kortekaas and Dawson (2007) described and interpreted the lithostratigraphy of Martinhal infill that includes the AD 1755 tsunami

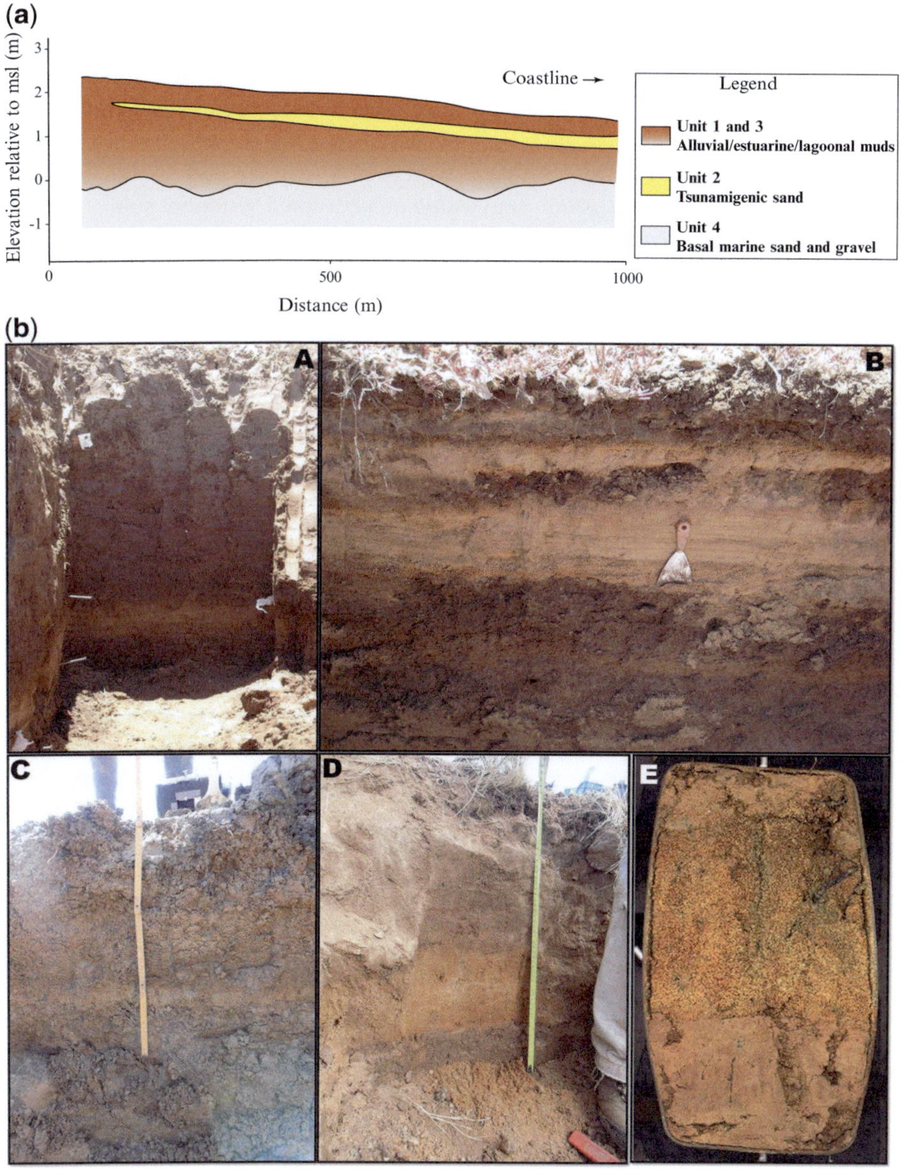

Fig. 2.4 (**a**) Schematic cross-shore profile showing major stratigraphic units characterizing the Late Holocene infill of the studied lowlands. (**b**) Sedimentological evidences of the AD1755 in lowlands of the Algarve. *A* Boca do Rio. *B* Martinhal. *C* Alcantarilha. *D* Salgados. *E* box core from Salgados. The vertical scale is app. 1 m for A, 0.30 m for E and app. 0.5 m for all the other images. Please note the contrasting texture of the sandy deposits and the low-energy framing (muddy) sediments. Please also note, that only in Martinhal the top of the sequence presents more than one sandy layer (for details please see discussion below)

deposit which extends across most of the lowland. This event layer marks a boundary between the pre and post sedimentation pattern, more evident in the seaward sector of the lowland. After 1755 evidence of storm activity is more frequent, translated in thin sandy overwash layers of small inland extension. The tsunami and storm deposits show similar grain size characteristics but distinctive contents in benthic foraminifera. The tsunami sand is generally coarser and less sorted than sand from the beach/dune system, suggesting an additional sediment source, possibly from the nearshore (e.g. Kortekaas 2002; Oliveira 2009).

In Barranco and Furnas lowlands, Costa et al. (2011) observed scattered cobbles and boulders (0.3–1 m a-axis) showing evidence of marine bioerosion and colonization by endolithic fauna indicative of an origin at a minimum depth of 5 m, that were transported inland a few 100 m into the alluvial plain. The radiocarbon dates obtained from *in situ* endolithic bivalve shells of *Petricola litophaga* (240 ± 40 BP and 360 ± 40 BP) suggests that boulder transport and deposition are compatible with the AD 1755 tsunami (Costa et al. 2011). Furthermore, in Barranco, a thin marine sand layer with shells associated with the boulder deposit has been detected embedded within the alluvial mud about 0.25 m beneath the surface and extending more than 300 m inland. One radiocarbon dating of 340 ± 30 BP obtained from a whole gastropod shell (*Cabestana* sp.) embedded in the sand is compatible with the above-cited dating results and with the same inundation. Grain size characteristics of this unit are similar to present day beach and foredune sands, suggesting these environments as the main sediment source (Oliveira 2009).

Dawson et al. (1995), Hindson et al. (1996) and Hindson and Andrade (1999) identified a coarse clastic layer attributed to the AD 1755 tsunami, within the Holocene infilling of Boca do Rio lowland, approximately 0.8 m beneath the surface, within alluvial muds. Eyewitnesses described this inundation and its effects in the destruction of a coastal dune at the back beach (Lopes 1841): "*In the day of the earthquake, the sea invaded the fresh water creek that outlets into the sea, for more than ½ league with a water height of 10–12 varas* [11–13 m] *destroying some large sand mounds* [dunes] *and carrying along 50 of the heaviest anchors more than ¼ league inland. The backwash uncovered great and noble buildings in the beach, next to the coastline, of which no memory existed*".

According to Dawson et al. (1995) and Hindson and Andrade (1999) the tsunamigenic unit marks a distinct sedimentological and micropalaeontological break from the deposits enclosing it. They noted that the lowermost section of the layer appears to have been deposited from a highly turbulent water mass, which was able to transport gravel-sized limestone clasts as well as rip-up mud clasts eroded from the underlying estuarine soft material. The water mass rapidly lost energy as it progressed inland, leading to the deposition of predominantly shell-rich sand, silt and clay particles, up to about 1000 m inland. The provenance study of Costa et al. (2012b) suggests the dune, and to a lesser extent the beach, as the most likely sediment sources of this tsunami layer.

The inundation of Alcantarilha lowland by the AD 1755 tsunami is described in historical records (Lopes 1841), which translate as: "*In…Armação* [Alcantarilha lowland], *located in the beach ¼ of league from another village named Pera, the sea left one house standing in the earthquake; it rushed more than ½ league inland, flooding everything, leaving salt water lakes in the lowlands, creating islands and drowning 84 people (…)*".

Costa et al. (2012a) characterized this tsunami deposit within Late Holocene lagoonal sediments of Lagoa dos Salgados, using many of the stratigraphical, textural, palaeoecological and compositional diagnostic criteria used to recognize this type of deposit elsewhere. Furthermore, age constraining using ^{210}Pb, ^{137}Cs and ^{14}C methods yielded results that were mutually consistent with the AD 1755 event (see Costa et al. 2012a for details on dating). Microtextural studies revealed a strong similarity between the dune and the tsunamigenic grains, suggesting the former as the likely source of the tsunami deposit (Costa et al. 2012a).

In Alcantarilha lowland, immediately to the west of Salgados, Dinis et al. (2010) refer a sandy overwash fan attached to the leeward toe of the dune ridge east of the inlet. The fan thins and wedges out inland into the adjacent alluvial plain sediments. The authors attribute its origin to localized overwash and erosion of the dune tip by the AD 1755 tsunami (Fig. 2.3); inedited data based on ^{210}Pb and ^{137}Cs profiles confirm this chronology.

In summary, the nature and particle size of tsunami sediments largely depend on the sediment available for transport, which in turn depends on local geology and geomorphological setting. Rocky coastlines tend to produce boulder deposits and depositional coasts (e.g. dune fields, beaches) tend to generate sandy deposits.

Baptista and Miranda (2009) used solely documentary records to argue that besides the AD 1755 event, other seismic-induced tsunamis (e.g. 60 BC and AD 382, also sourced offshore Iberia) affected the Portuguese coastline. However, to the best of our knowledge, only the AD 1755 has been unequivocally detected in the Portuguese onshore geological record so far. This is probably due to a clear textural contrast between host and tsunami sediment found in low energy lowlands, which facilitates its association to an abrupt high energy marine inundation. The textural contrast is unlikely in sediment pre-dating the establishment of barriers.

2.2.4 Geomorphological Imprints and Post-impact Recovery

The evolution of Algarve lowlands impacted by the AD 1755 event can only be based on geological and geomorphological features, added by limited ichnographic information and historical descriptions. From the analysis conducted, including aerial imagery coupled with LiDAR data, one can conclude that the five studied sites did not fully recover pre-event morphology. In the case of Martinhal, this statement is supported by the contrast in storm penetration ability into the back-barrier wetland before and after the tsunami. Sedimentation records this change in the frequency of sandy layers in the upper unit of the alluvial plain stratigraphic column. This is interpreted as caused by a substantial decrease in the sheltering efficiency of the beach-dune system in the last 250 years. It is reasonable to assume that the tsunami inundation transferred a large volume of sand into the eighteenth century lagoon, which remained entrapped there since then, contributing to its terrestrialization. The low rate of sand supply to the coastal cell and the volume of sand abruptly bypassed by that single event support the reasoning that, once eroded by the tsunami waves, the barrier was not able to fully recover, until present.

Boca do Rio illustrates another case of failure of the geomorphologic recovery of the coastal system. As in Martinhal, this coast is sediment starved and the ablation of the large foredune sitting in the eighteenth century at the seaward edge of the lowland by the AD 1755 tsunami (Lopes 1841) permanently changed the local morphology. Moreover, the remnants of the dune exhibits approximately 2–4 m scarps produced by the inundation (Oliveira 2009) confirming its destructive character and the incapability of the system to recover.

In the Alcantarilha-Salgados region the pre-event elevations of most of the beach/dune system were higher than the tsunami wave height (Costa et al. 2012a). In this case, the dune robustness provided a necessary buffer for tsunami wave overtopping and overwash. However, in the western tip of the barrier, where the dune is lower, the emplacement of a washover fan (Fig. 2.3) is attributed to dune overtopping by the AD 1755 tsunami wave (Dinis et al. 2010). Besides this, one interesting post-event geomorphological feature observed in Alcantarilha- Salgados landward strip of the sand barrier, are parabolic dunes (Fig. 2.3). Similar features have been previously described in the literature in areas affected by tsunamis (Goff et al. 2009 and Atwater et al. 2013) in relation with post-event remobilization of sand deposited landward, in environments characterized by low moisture content. Interestingly, the feature described in the latter work accumulated in the seaward sector of the coastal profile.

The analysis of aerial imagery, maps and DEMs, revealed also a number of scarps affecting the dunes. The inner scarps, with 30–70 m length and 2–4 m height, were not obliterated nor retreated since at least 1947 (earliest aerial photographic record). Considering that during the last century extreme storms that affected this coast were not capable of changing or eroding these scarps, its origin can be attributed to a unique extreme event, most likely the AD 1755 tsunami. Nowadays, the parabolic dunes are fixed by vegetation, which implies a reduction of aeolian sand remobilization. One can hypothesize that the present-day geomorphology is inherited from syn-event erosion and later wind reshaping of a poorly or non-vegetated surface, temporarily increasing sediment availability for aeolian transport.

The results presented here are in contrast with recent work conducted in the aftermath of the 2004 Indian Ocean tsunami and the 2011 Japan tsunami that focused in post-tsunami recovery of coastal systems. In fact, in Lhok Nga (Indonesia), approximately 1 year after the tsunami impact, the coastal system had apparently recovered its typical morphology and sedimentary profile (cf. Paris et al. 2009 and Catalán et al. 2014).

Comparison between the cases presented in this work and the one above suggests that the Portuguese sites present a large recovery time to extreme marine events of this magnitude and this is caused by differences in sediment input, climatic conditions and hydrodynamic regime.

2.2.5 Summary

Tsunami imprints are usually associated with either erosional or depositional features. The latter typically consist of a peculiar sedimentological unit embedded in coastal stratigraphy. These units have been frequently identified as sandy deposits in backshore low energy environments (e.g. alluvial plains, estuarine/lagoonal basins). The AD 1755 left exotic sediment in several lowlands of the Algarve. In this chapter these materials were briefly described and interpreted. In terms of erosional features in soft sediments, they are mostly associated with scours and scarps. This work describes the few erosive geomorphological features that have been recognized in the study area. Furthermore, the response of the studied coastal systems is discussed having in consideration that full recovery is conditioned by sediment supply. In fact, systems with moderate or poor sediment abundance will take long periods to recover or recovery simply does not occur. This seems to be the case of the Portuguese studied sites which is in contrast with Indonesian, or Chilean sites (e.g. Paris et al. 2009 and Catalán et al. 2014).

In sand barriers, sediment availability for post-impact reworking is conditioned by several factors, namely vegetation cover and moisture content. After tsunami inundation, the normal scenario is the partial or total disappearance of the vegetation cover (by erosion, burial or extensive contact with salt-water). Climatic conditions after the tsunami event (i.e. moisture content and wind) could contribute to facilitate sediment availability. In dryer (or lower moisture content) environments, added by vegetation rarefaction, the remobilization of sand is more common, especially with persistent onshore winds which favors the formation of parabolic dune fields. In a wetter (or higher moisture content) environment the sediment is stabilized and is not available for movement, which results in harder conditions in the recovery from tsunami changes. In addition, part of the sediment eroded by the tsunami may have been made available to be reworked by onshore winds, reshaping the topographic profile resultant from the event and contributes to form the parabolic dune field. Both scenarios show none to limited recovery to its pre-event geomorphological configuration.

Figure 2.5 illustrates a conceptual model schematizing the geomorphological response of Portuguese coastal systems to a major tsunami event. Figure 2.5a, b, c, represents a context where the barrier height is lower than the tsunami height (e.g. Martinhal, Barranco, Furnas, Boca do Rio and near-inlet areas of Alcantarilha and Salgados). These sediment-starved areas will present depositional imprints inland and erosional geomorphological features still identifiable (in the field) due to low recovery rates. Figure 2.5d, e, f illustrates contexts where barrier is higher than the tsunami wave (e.g. coastal ribbon between Salgados and Alcantarilha lowlands). This type of setting will not produce inland deposits and the tsunami-related geomorphological features will have essentially an erosive character and locate in the seaward part of the barrier (deposits could be formed solely in the nearshore). Its recovery is potentially easier than in the previous scenario but will depend on sediment input (that is determined by the regional/local geological, hydrodynamic and climatic contexts).

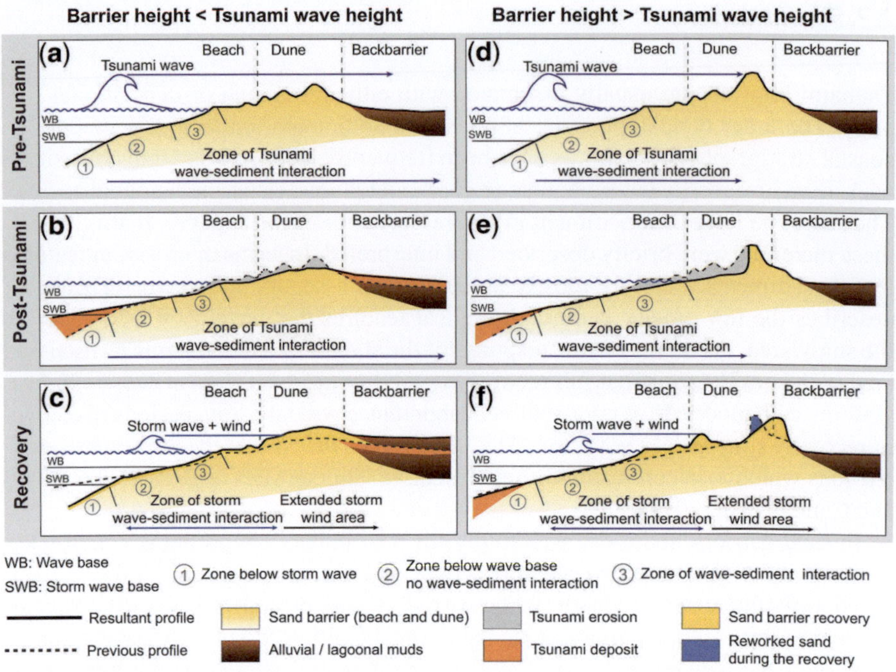

Fig. 2.5 Conceptual model of sand barrier recovery after a tsunami event assuming moderate sediment availability. Two possible scenarios during a tsunami event are consider: barrier height lower (**a**) and higher (**d**) than the tsunami wave height. The resultant profile after a tsunami event for each possible scenario is presented in (**b**) and (**e**). The recovery after a short geologic time is illustrated in (**c**) and (**f**)

In conclusion, the study of tsunami imprints is highly conditioned by local factors (e.g. setting, sediment supply, climate, hydrodynamic regime) that are also crucial for post-event recovery.

Acknowledgements MA Oliveira and PJM Costa benefited from FCT (Portuguese Science Foundation) PhD and Post-Doc Scholarships. R. González-Villanueva benefited from an IDL Post-Doc Fellowship and Xunta de Galicia (PlanI2C-ED481B 2014/132-0) Post-Doc Fellowship.

References

Andrade C (1990) O ambiente de barreira da Ria Formosa (Algarve, Portugal). PhD dissertation. University of Lisbon, Portugal, 645 pp

Andrade C (1992) Tsunami generated forms in the Algarve barrier islands (South Portugal). Sci Tsunami Haz 10:21–34

Andrade C, Andrade AM, Kortekaas S, Dawson A (1997) Sedimentological traces of tsunamigenic overwash of the Martinhal lowland (Western Algarve – Portugal). In: Proceedings Seminar Zona Costeira do Algarve, Faro, 10–12 July 1997, Eurocoast-Portugal, pp 11–18

Atwater BF (1987) Evidence for great Holocene earthquakes along the Outer Coast of Washington State. Science 236:942–944

Atwater BF, Cisternas M, Yulianto E, Prendergast AL, Jankaew K, Eipert AA, Fernando WIS, Tejakusuma I, Schiappacasse I, Sawai Y (2013) The 1960 tsunami on beach-ridge plains near Maullín, Chile: landward descent, renewed breaches, aggraded fans, multiple predecessors. Andean Geol 40:393–418

Atwater BF, Fuentes Z, Halley RB, Ten Brink US, Tuttle MP (2014) Effects of 2010 Hurricane Earl amidst geologic evidence for greater overwash at Anegada, British Virgin Islands. Adv Geosci 38:21–30

Bahlburg H, Spiske M (2011) Sedimentology of tsunami inflow and backflow deposits: key differences revealed in a modern example. Sedimentology 59:1063–1086

Baptista MA, Miranda JM (2009) Revision of the Portuguese catalog of tsunamis. Nat Hazards Earth Syst Sci 9:25–42

Catalán PA, Cienfuegos R, Villagrán M (2014) Perspectives on the long-term equilibrium of a wave dominated coastal zone affected by tsunamis: the case of Central Chile. J Coast Res 71:55–61

Chagué-Goff C, Schneider JL, Goff JR, Dominey-Howes D, Strotz L (2011) Expanding the proxy toolkit to help identify past events – lessons from the 2004 Indian Ocean Tsunami and the 2009 South Pacific Tsunami. Earth Sci Rev 107:107–122

Costa PJM, Andrade C, Freitas MC, Oliveira MA, da Silva CM, Omira R, Taborda R, Baptista MA, Dawson AG (2011) Boulder deposition during major tsunami events. Earth Surf Process Landf 36:2054–2068

Costa PJM, Andrade C, Freitas MC, Oliveira MA, Lopes V, Dawson AG, Moreno J, Fatela F, Jouanneau J-M (2012a) A tsunami record in the sedimentary archive of the central Algarve coast, Portugal: characterizing sediment, reconstructing sources and inundation paths. The Holocene 22(8):899–914

Costa PJM, Andrade C, Dawson AG, Mahaney WC, Paris R, Freitas MC, Taborda R (2012b) Microtextural characteristics of quartz grains transported and deposited by tsunamis and storms. Sediment Geol 275–276:55–69

Dawson AG (1994) Geomorphological effects of tsunami run-up and backwash. Geomorphology 10:83–94

Dawson AG, Stewart I (2007) Tsunami deposits in the geological record. Sediment Geol 200:166–183

Dawson AG, Long D, Smith DES (1988) The Storegga slides: evidence from eastern Scotland for a possible tsunami. Mar Geol 82:271–276

Dawson AG, Hindson R, Andrade C, Freitas C, Parish R, Bateman M (1995) Tsunami sedimentation associated with the Lisbon earthquake of 1 November AD 1755: Boca do Rio, Algarve, Portugal. The Holocene 5(2):209–215

Dinis J, Andrade C, Oliveira MA, Freitas MC, Cunha PP, Martins A, Costa P (2010) Geomorphological constraining of tsunami (?) run-up in the Alcantarilha coastal lowland (central Algarve, Portugal). In: Proceedings Iberian coastal holocene paleoenvironmental evolution 'Coastal Hope 2010', Lisbon, pp 40–41

Goff JR (2008) The New Zealand Palaeotsunami database. NIWA technical report 131, ISSN 1174–2631, 24 pp + Appendix

Goff J, Lane E, Arnold J (2009) The tsunami geomorphology of coastal dunes. Nat Hazards Earth Syst Sci 9(3):847–854

Hindson RA, Andrade C (1999) Sedimentation and hydrodynamic processes associated with the tsunami generated by the 1755 Lisbon earthquake. Quat Int 56:27–38

Hindson RA, Andrade C, Dawson AG (1996) Sedimentary processes associated with the tsunami generated by the 1755 Lisbon earthquake on the Algarve coast, Portugal. Phys Chem Earth 21:57–63

Hindson R, Andrade C, Parish R (1999) A microfaunal and sedimentary record of environmental change within the late Holocene sediments of Boca do Rio (Algarve, Portugal). Geol Mijnb 77:311–321

Kain CL, Gomez C, Hart DE, Wassmer P, Goff J, Starheim C (2014) Assessing topographic controls on flow direction in washover deposits using measurements of magnetic fabric. Mar Geol 350:16–26

Kortekaas S (2002) Tsunamis, storms and earthquakes: distinguishing coastal flooding events. Coventry, PhD dissertation. Coventry University, United Kingdom, 171 pp

Kortekaas S, Dawson AG (2007) Distinguishing tsunami and storm deposits: an example from Martinhal, SW Portugal. Sediment Geol 200(3–4):208–221

Lopes JBS (1841) Corografia ou Memória Económica, Estatística e Topográfica do Reino do Algarve. Academia R. das Sciencias de Lisboa, Lisboa, Portugal, 528 pp (in Portuguese)

Oliveira MA (2009) Influência da geomorfologia local na preservação de assinaturas sedimentares de eventos de alta energia no Algarve ocidental. MSc dissertation, University of Lisbon, Portugal, 178 pp

Oliveira MA, Andrade C, Freitas MC, Costa PJ (2009) Modeling volume transfer between beach-foredune and the backshore by the 1755 Lisbon tsunami at Boca do Rio lowland, Algarve (Portugal). J Coast Res SI56(2):1547–1551

Paris R, Wassmer P, Sartohadi J, Lavigne F, Barthomeuf B, Desgages E, Grancher D, Baumert P, Vautier F, Brunstein D, Gomez CH (2009) Tsunamis as geomorphic crises: lessons from the December 26, 2004 tsunami in Lhok Nga, West Banda Aceh (Sumatra, Indonesia). Geomorphology 104:59–72

Pires HO (1985) Alguns aspectos do clima de agitação marítima de interesse para a navegação na costa de Portugal. Instituto Nacional de Metereologia e Geofísica, Lisboa, Portugal, 30pp (in Portuguese)

Chapter 3
Ecosystem-Based Tsunami Disaster Risk Reduction in Indonesian Coastal Areas

Eko Rudianto, Abdul Muhari, Kenji Harada, Hideo Matsutomi, Hendra Yusran Siry, Enggar Sadtopo, and Widjo Kongko

Abstract A healthy natural coastal ecosystem can function as one of the components in reducing potential risk of coastal disasters. The impacts of tsunamis, storm surges and coastal erosions can be reduced at a certain limit by the existence of coastal forest and dunes. In Indonesia, tsunami occurs once twice a year in average. It means, tsunami hit quiet frequently even though the return period in a specific location mostly is several tens to hundred years. To reduce potential impacts of tsunamis in coastal area, construction and rehabilitation of coastal forest is one of the main efforts. The existence of a healthy coastal forest not only provides a suitable protection for high frequency but relatively minor to medium scale tsunamis, but also promotes economic activity based on eco-tourism that will ensure the sustainability of the coastal forest maintenance in the later phase. This paper aims to describe milestones of tsunami mitigation by using greenbelt in Indonesia. Conception, tsunami hazards assessment, challenges and lessons learnt in applying tsunami mitigation by using greenbelt are described so the initiative can be replicated in other tsunami prone areas.

Keywords Tsunami hazards • Tsunami model • Coastal forest

E. Rudianto • A. Muhari (✉) • H.Y. Siry • E. Sadtopo
Directorate for Coastal and Ocean, Ministry of Marine Affairs and Fisheries, Republic of Indonesia, Jakarta, Indonesia
e-mail: abdul.muhari@gmail.com

K. Harada
Center for Integrated Research and Education of Natural Hazards, Shizuoka University, Shizuoka, Japan

H. Matsutomi
Department of Civil Engineering, Akita University, Akita, Japan

W. Kongko
Coastal Dynamics Research Center, BPPT, Jakarta, Indonesia

V. Santiago-Fandiño et al. (eds.), *Tsunamis and Earthquakes in Coastal Environments*, Coastal Research Library 14, DOI 10.1007/978-3-319-28528-3_3

3.1 Introduction

3.1.1 Tsunami Hazards in Indonesia

Indonesia has been identified as one of the most prone countries to tsunami disasters in the world where about 5.5 million of the total 205 million people who live in the area at risk are within the potential tsunami inundation zone (UNISDR 2013; Lovholt et al. 2014; Pardede 2014). From the official tsunami risk map for Indonesia established by the National Agency for Disaster Management (BNPB 2011), it can be seen that west coast of Sumatra, south coast of Java, south Sulawesi, Moluccas and Papua Province coasts have potential tsunami risk from medium to high level (Fig. 3.1).

Broad and detailed tsunami hazards and risks assessments focusing in Sumatra and Java Island has been published by Strunz et al. (2011). Some other specific studies covering these areas have also been published (Gayer et al. 2010; Taubenbock et al. 2009). As West Sumatra and Bengkulu Province in Sumatra Island are considered to be the next highest tsunami risk areas, a number of studies have been undertaken (Muhari et al. 2011; Borrero et al. 2006). In Padang City (capital of West Sumatra Province), Muhari et al. (2011) and Imamura et al. (2012) identified the potential tsunami inundation concluding that it may reach up to 3 km inland and is potentially inundating an area around 21.46 km^2. Borero et al. (2006) modeled several earthquake scenarios in Bengkulu City (Bengkulu Province) to estimate the tsunami potential inundation area concluding that the inundation distance may reach from 0.6 km to 6 km from the coastline.

In Java and Nusa Tenggara, aside of the three tsunami events namely the 1977 Sumba tsunami (Gusman et al. 2009), 1994 East Java Tsunami (Tsuji et al. 2012) and 2006 West Java tsunami (e.g. Fujii and Satake 2006; Fritz et al. 2007), some other potential earthquake and tsunami in the south coast of Java have also been assessed (Hanifa et al. 2014; Okal and Synolakis 2008).

It is important to note that the high level of tsunami hazard and risk potential is not limited only on the Sumatra and Java Island alone as similar situation has also been identified in the eastern part of Indonesia (Okal et al. 2011; Lovholt et al. 2012; Diposaptono et al. 2013).

3.1.2 Approaches in Reducing Potential Tsunami Risks

In general, the tsunami disaster mitigation policy in Indonesia is directed to minimize risks to people's lives and economy as well as to natural resources, environment, livelihoods and the society (Muhari et al. 2007). In light of this, a number of technical actions such as the construction of coastal forest amongst other measures have already been developed (Muhari et al. 2010, 2012), which runs in parallel with social and cultural components (Spahn et al. 2010).

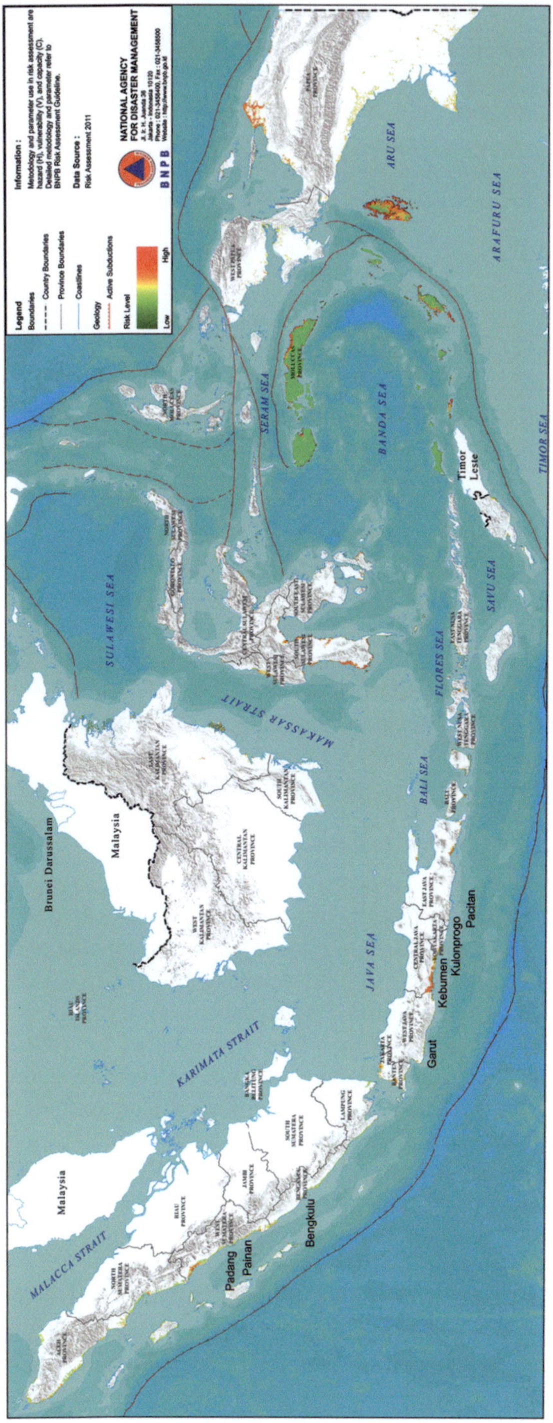

Fig. 3.1 Indonesia National Tsunami Risk Map. National Agency for Disaster Management; Indonesia

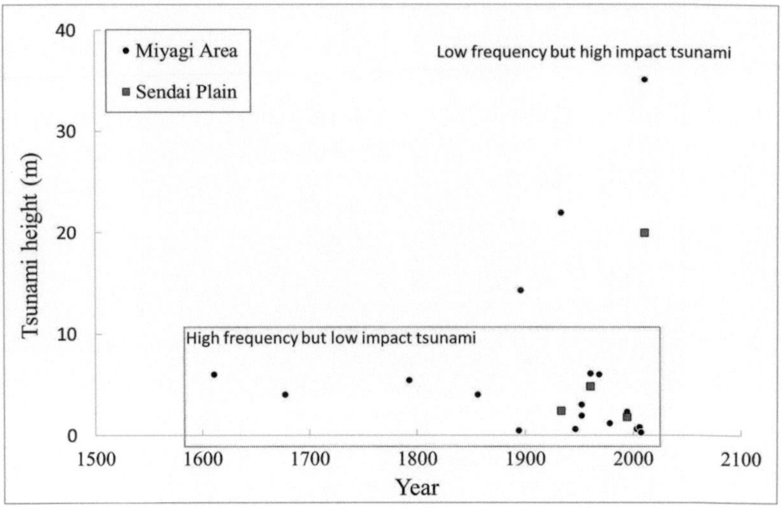

Fig. 3.2 Historical tsunami height in Miyagi coasts (central part of Tohoku coast, Japan) and in Sendai City that shows an exceptional height for the case of 2011 Tohoku tsunami compared to the other cases in the region that have less than 10 m maximum tsunami height

As result of the March 2011 tsunami in Japan, most of coastal forests along the Tohoku coast perished giving rise to strong skepticism on their capabilities to reduce tsunami impacts. However, considering the fact that the maximum tsunami height reached up to 32 m in Miyagi Prefecture and about 18 m in the Sendai plain, this event can be considered as an exceptional one; moreover, the likelihood of tsunamis with this magnitude only occur one in several hundred to thousand years. On the other hand, the historical record shows that the more frequent tsunamis with a maximum height less than 10 m occur within 50–100 years return period (Muhari et al. 2014, Fig. 3.2). Considering this fact, coastal forests could prove to be a very helpful approach to protect vulnerable and exposed coastlines in this part of Japan.

Similar situation was observed in Indonesia. The maximum tsunami height in the case of 2004 Indian Ocean tsunami is also an exception. By looking at the post tsunami survey records obtained from recent events such as the 2006 West Java tsunami (Fritz et al. 2007), 2007 Bengkulu tsunami (Borrero et al. 2009) and the 2010 Mentawai tsunami (Satake et al. 2013) the observed tsunami heights are roughly about 5 m although in certain places depending on the shoreline configuration and or the possibility of the additional local tsunami source such as landslide, the maximum height can exceed 15 m. Considering this situation, the government of Indonesia believes that coastal forests or greenbelts are amongst the most important protecting methods to reduce the impact of tsunamis in vulnerable coastlines.

A tsunami hazard assessment study aiming at the development and or enhancement of the options to mitigate future tsunami disasters to occur in Sumatra and Java Islands has been undertaken. In this Chapter, the use of coastal forest as one of the mitigation options to reduce onshore tsunami impacts is discussed. The role of coastal forest to reduce tsunami impacts based upon scientific principles, the expected milestones towards their growth in Indonesia in parallel with a tsunami hazard assessment undertaken in various regions such as Sumatra and Java Island, lessons learned and the possible way ahead are explained and the chapter's conclusions are finally presented aiming at promoting the use of a natural barrier such as the coastal forest to reduce and mitigate tsunami impacts, which could prove to be of particular importance for developing countries where sophisticated and high cost structural protective structures and other countermeasures may not be affordable.

3.2 Scientific Considerations for the Establishment of Coastal Forests for Tsunami Mitigation

3.2.1 Experiences from the 2004 Indian Ocean Tsunami: Damage on Mangrove Forest in Indonesia

The 2004 Indian Ocean tsunami damage on mangrove forest was observed in Indonesia and Thailand amongst other countries. Yanagisawa et al. (2010) conducted field survey in 190 ha of mangrove forest in Banda Aceh, Indonesia to identify the damage characteristics and its relationship with tsunami parameters. They found that more than 50 % of mangrove trees with 20–25 cm stem diameter could survive a tsunami with less than 6–7 m in height, but could not stand it when the tsunami height were 7–9 m or higher. By using numerical model, they also demonstrated that 500 m mangrove forest can effectively reduce the maximum tsunami inundation up to 38 %, and the hydrodynamic force behind the mangrove forest up to 70 % if the tsunami is about 3 m height when reaching the mangrove forest. Considering the study, one might say that the capability of mangrove forest to reduce hydrodynamic force is higher than its capabilities to reduce tsunami inundation depth.

Another important conclusion in the study is that the reduction rate of tsunami impact by the mangrove forest relates to age; a 10 years old mangrove forest will successfully reduce the impact of a 3 m tsunami height event, while 20–30 years old will optimally reduce a 4 m tsunami. Although one cannot expect much impact reduction capability of mangrove forest for tsunamis beyond 4 m height, the results stresses the need to maintain coastal mangrove forest across generations to ensure their mitigating effectiveness.

Ohira et al. (2012) studied the damping performance of coastal pine tree forest considering several widths for a 4 m maximum tsunami height by using numerical models in Yogyakarta, Indonesia. The study demonstrated that a 100 m width coastal pine tree (*Casuarina equisetifolia*) forest with a 4 trees/100 m^2 density and 0.2 m trunk diameter can reduce the inundation flux, inundation depth and inundation area by 17.6, 7.0 and 5.7 % respectively. It means, similar to the case of mangrove forest, the capability of pine tree forest to reduce the inundation flux is more significant compared to its influence in reducing inundation depth.

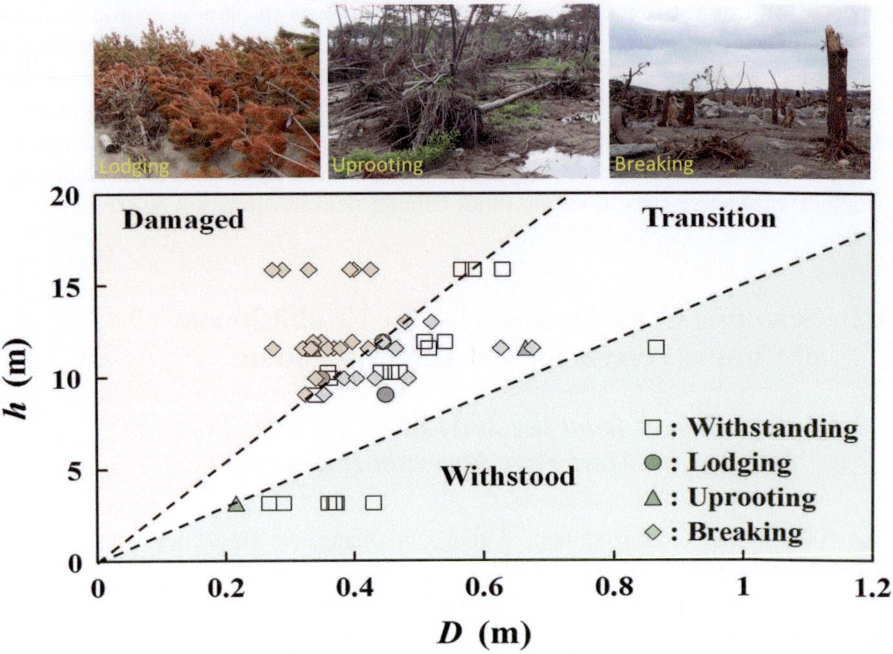

Fig. 3.3 Relationship between the trunk diameter D, inundation depth h and damage type of coastal black pine tree in the 2011 Tohoku tsunami (Matsutomi et al. 2012)

3.2.2 Experiences in the 2011 East Japan Earthquake-Tsunami

Matsutomi et al. (2012) examined damage condition of coastal black pine trees (*Pinus thunbergii*) after the 2011 East Japan Tsunami through field observations and tests. He concluded that the sustained damage could be classified in three different types i.e. lodging, uprooting and breaking as seen in Fig. 3.3. Matsutomi et al. (2011) also observed that a similar tendency occurred in tropical pine trees (*C. equisetifolia*) in Indonesia; moreover that the breaking also affected lodging and uprooting.

Figure 3.3 shows that (1) as the increment on the ratio of the inundation depth to the trunk diameter –the diameter of the tree before the first branch, becomes larger the tendency of breaking becomes stronger, which is widely accepted since similar tendency was observed in the case of coastal pine tree in Indonesia as observed in Matsutomi et al. (2011), which the breaking data extend over both domains lodging and uprooting. (2) the uprooting force tends to be smaller than that of lodging.

Although a previous study done by Shuto (1987) suggests that coastal forests are only effective to reduce the impact of a tsunami with less than 5 m height, Matsutomi et al (2012) observed their importance in reducing inundation depth in the case of tsunami heights ranging from 3 to 6 m or more in the southern part of Sendai coast. Furthermore, he also concluded from the field observations that a coastal forest of

220 m width has a potential capability to reduce tsunami inundation depth and inundation height up to 25 %.

From the above studies, it can be summarized that (1) coastal forest has the potential capability to reduce tsunami impact for small to moderate tsunami (up to ~5 m height), although in some cases coastal forest was observed to be able to reduce tsunami height in a larger magnitude; (2) the main function of coastal forest (mangrove and pine tree forest) probably is not to reduce tsunami inundation depth but rather to reduce the hydrodynamic force. In line of this, coastal forest might significantly reduce the tsunami velocity, thus minimizing the impact in areas behind the forest belt. These are the underlying consideration for the continuation of the coastal forest development in Indonesia.

3.3 Milestone of Tsunami Mitigation by Using Green Belt in Indonesia

3.3.1 General

Indonesia began their tsunami disaster mitigation activities using coastal vegetation in 2008. The initiative started in Pacitan, East Java where historical records of tsunami events indicate a seismic gap between the tsunami-earthquake in 1994 in eastern Java and the one in 2006 in West Java. Muhari et al. (2012) simulated the tsunami in Pacitan found that a M8.5 earthquake can generate with maximum and average inundation depth of 10.8 m and ~4 m respectively (Fig. 3.4a, b).

Coastal forest planting in Pacitan has been undertaken in several stages during the past 6 years. In Pacitan the coverage has now reached 14 ha with as many as 42,500 planted *C. equisetifolia* from which 23,000 were planted in 8 ha from 2008 to 2010 and 19,500 covering 6.9 ha in the period 2011–2014 (Fig. 3.4c–e).

In Sumatra, construction of coastal forests with species of *Casuarina spp* was carried out in two provinces namely Bengkulu and West Sumatra following the result of the 2007 Southwestern Sumatra tsunami where the coast of these two Provinces were hit by tsunami with a maximum height of 4 m (Borrero et al. 2009). Planting coastal vegetation in these two provinces started in 2013. In West Sumatra, the construction was carried out in six districts which cover all the coastal districts along the coast. In Bengkulu, the construction of coastal forests was undertaken in five districts hence also covering the entire coast of this Province with the exception of Bengkulu City.

In Bengkulu Province, the coastal forests equal 44 ha with as much as 79,238 stems of *C. equisetifolia* whereas in the province of West Sumatra the total reached 52.5 ha totaling 58,376 stems of the same species (Table 3.1). The establishment and development of the coastal forest has been done in separate locations as land availability for cultivation varies depending on the particular site in a given district, therefore at one point the total amount of coastal forest area may not be too large. For instance, in West Sumatra and Bengkulu provinces planting have been done in 34 points covering an area between 1.5 and 16 ha.

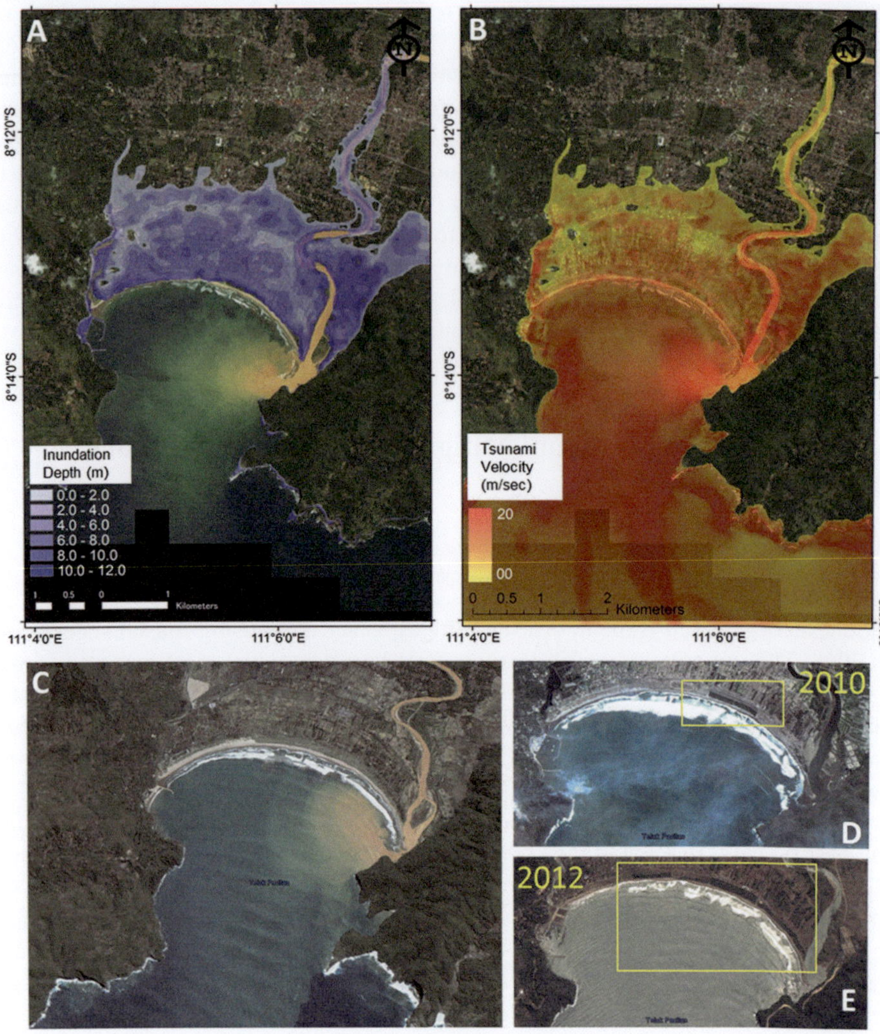

Fig. 3.4 (**a**) The result of the numerical model for inundation depth and (**b**) tsunami velocity in Pacitan. (**c**) Pacitan beach before construction of coastal forest, (**d**) Coastal forest section 1 built on 2008–2010 and (**e**) the current condition of coastal forest in Pacitan

3.3.2 The Growth Rate of Pine Trees in Indonesia

The most crucial time in planting coastal forest is the time between planting seeds and their adaptation to the new environment. During this particular period the plants are in a high vulnerable condition due to weather factors, other environmental conditions as well as human or animal interference. This critical period last until the plant becomes 1 year old.

During this phase the plants should be monitored to measure growth and survival rates. Moreover, plant height (*H*), trunk diameter (*D*), height of the first branch from

Table 3.1 Summary of the greenbelt construction in Indonesia

No	Location				
	District	Province	Year	Area (Ha)	Total number of stems
1	Pacitan	East Java	2008	8.06	23,000
2	Pacitan	East Java	2011	6.91	19,500
3	Pasaman Barat	West Sumatra	2013	12.08	13,420
4	Agam	West Sumatra	2013	8.64	9597
5	Padang Pariaman	West Sumatra	2013	9.82	10,908
6	Pariaman	West Sumatra	2013	8.43	9370
7	Padang	West Sumatra	2013	2.80	3114
8	Pesisir Selatan	West Sumatra	2013	10.77	11,967
9	Muko Muko	Bengkulu	2013	10.77	16,529
10	Bengkulu Tengah	Bengkulu	2013	16.92	40,756
11	Seluma	Bengkulu	2013	7.30	8111
12	Bengkulu Selatan	Bengkulu	2013	7.64	10,192
13	Kaur	Bengkulu	2013	1.46	3650
14	Kulonprogo	Yogyakarta	2015	20	30,000
15	Kebumen	Central Java	2015	20	28,000
16	Garut	West java	2015	5	5000

the ground (trunk height) and branches width (B) should also be measured periodically (Fig. 3.5).

The above-mentioned parameters have also been very useful in numerical models to assess tsunami reduction due to the presence of coastal forest. Harada et al. (2011) measured them all in two local subspecies of *C. equisetifolia namely shrimp-* and *sea-casuarina* (local name: *cemara udang* and *cemara laut* respectively) in the Painan District (West Sumatra) and Pacitan District (East Java). These typical pine tree subspecies in Indonesia were planted in 2009 and 2010 meaning that the plants were between 1 and 2 years old at the time when the measurements were made. Field observation shows the relationship between plant height and trunk diameter at the time (Fig. 3.6). For *sea-casuarina*, trunk diameter up to 15 cm correlated with height up to 10 m and 30–40 cm correlated with plant height of 20–30 m.

For *shrimp-casuarina* trees, however, different results were observed. The maximum height of 1-year-old plant showed to be limited up to 10 m height tending not to grow vertically anymore despite the continuous increase in trunk diameter up to 35 cm.

Using the above mentioned field data (Harada et al. 2011) the trees growth rate curve has been calculated for the Indonesian case by using the Mitscherlich function as follows:

$$w = A\left(1-e^{-kt}\right) \tag{3.1}$$

where, w is the growth parameter (tree height, trunk diameter and etc.), A is the maximum value of 'w', k is growth speed and t is the time (tree age).

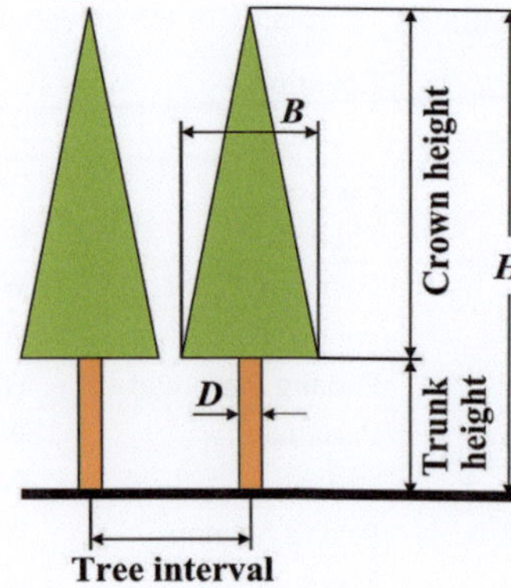

Fig. 3.5 Tree parameters to be measured during the monitoring of health and survival rate of tree (Matsutomi et al. 2011)

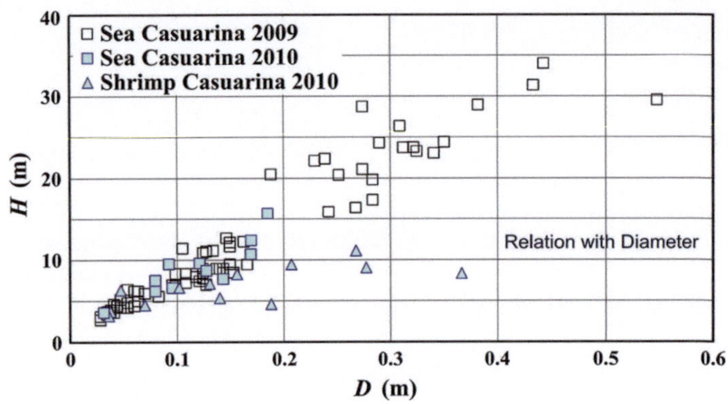

Fig. 3.6 Correlation between the height of pine tree and trunk diameter for Casuarina trees subspecies *Sea-Casuarina* and *Shrimp-Casuarina*

The regression results by using the above equation are given in Fig. 3.7 showing the relationship between age, diameter and tree height. In the case of *sea casuarinas* trunk's diameter in Indonesia, it will reach a maximum between ~30 and 40 cm diameter for 30-year-old plants, and height will reach its peak of about 30 m at that age too.

3.4 Numerical Modeling for Tsunami Reduction Impact

In order to continue the development of coastal forest in tsunami prone area in Indonesia, three areas for coastal forest development have been chosen in 2015 at the south coast of Java namely Yogyakarta, Central Java and West Java Province

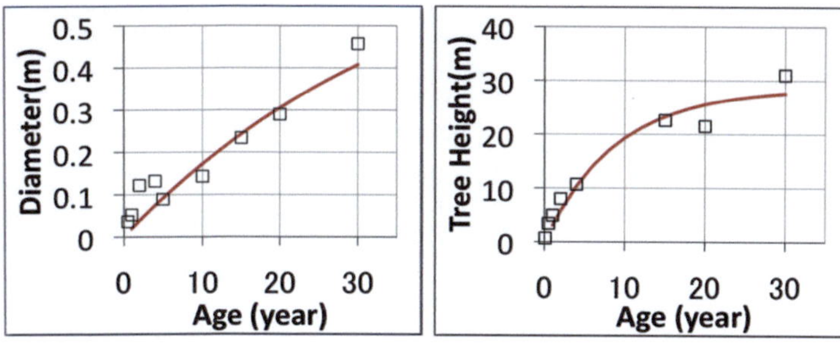

Fig. 3.7 Growth curves of Casuarina trees in Indonesian coast

Table 3.2 Growth rate parameters of *Casuarina equisetifolia* in the Indonesian coast

Tree age (year)	1	5	10	15	20	30
Tree height (m)	5.4	10.4	15.7	20.3	24.1	30.1
Diameter (m)	0.04	0.11	0.19	0.25	0.32	0.42
Height of branch (m)	0.97	1.87	2.83	3.65	4.34	5.42
Tree density (tree/m^2)	0.11	0.11	0.63	0.063	0.063	0.028
Forest width (m)	50	50	50	50	50	50

(Table 3.1). Before planting we conducted numerical exercises to determine the potential reduction of tsunami impact by using the aforementioned data of the growth rate function and taking the Kulonprogo District, Yogyakarta as a case study.

The tsunami propagation and inundation model has been simulated by using a fully validated model known as TUNAMI -Tohoku University Numerical Analysis Model for Investigation- (IOC-UNESCO 1997; Imamura 1996, 2009). The existence of coastal forest has been represented by means of expanding the bottom friction as per Ohira et al. (2012).

To integrate the information from the growth rate function (w), a simpler representation of the growth parameter is shown in Table 3.2. The parameters of 10 years old tree were selected to run the numerical model since the maximum growth rate occurred up to 10 years of age, thereafter the growth rate become slower. The selected forest width was 50 m to match with the existing actual condition in the Kulonprogo district.

The topography data for numerical simulation obtained from the field survey data of multi-line cross-section profiles from the lowest water level up to about 500 m perpendicular to the shoreline conducted by the Indonesian Ministry of Marine Affairs in 2014. This information has been combined with surveyed bathymetric data in order to create a 7 m cell size resolution for the area of coastal forest in Kulonprogo (inset box in Fig. 3.8). To run the tsunami simulation, an earthquake scenario is included by considering the work by Hanifa et al. (2014) and assuming a hypothetic earthquake with M8.3 occurs at about 180 km off the Kulonprogo coast (Fig. 3.8). The detailed earthquake parameters are given in Table 3.3.

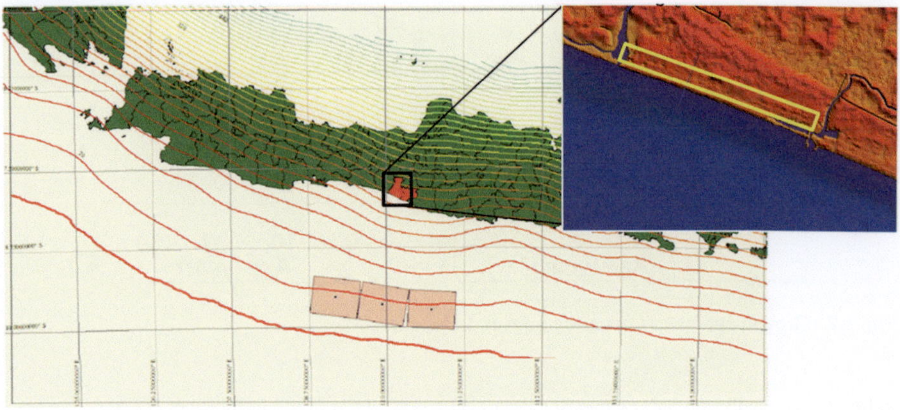

Fig. 3.8 Earthquake sources with M8.3 and inset box the location of the greenbelt construction area in Kulonprogo District

Table 3.3 Earthquake parameter for numerical simulation

No.			Mag.	Depth	Dislocation of sea bed					
Fault	Lon.	Lat.	(Mw)	(km)	Strike	Dip	Slip	L (km)	W (km)	rake (m)
1	109.93	−9.639	8	17	277	11	90	83	65	13
2	109.166	−9.524	8	17	280	14	90	83	65	13
3	110.715	−9.729	8	17	273	14	90	83	65	13

The result of the numerical analyses as is given in Fig. 3.9 indicating that the existence of 50 m width greenbelt is capable to reduce the maximum inundation depth up to 13.57 % and the maximum tsunami velocity up to 18.4 %. Hence, the coastal forest reduces the potential inundation area from 304.9 ha to 218.38 ha or about 28.37 % and thus make it is still reliable to be developed because there are many infrastructures to be built in Kulonprogo coast in the future.

3.5 Some Other Considerations, Best Practices and the Way Ahead

The positive impact of coastal forest development as shown in this chapter appears to be very high as also in addition to its primary function as a natural barrier to reduce the impact of tsunamis. The other indirect benefits of these structures provide tremendous gains to leeward communities behind the belt:

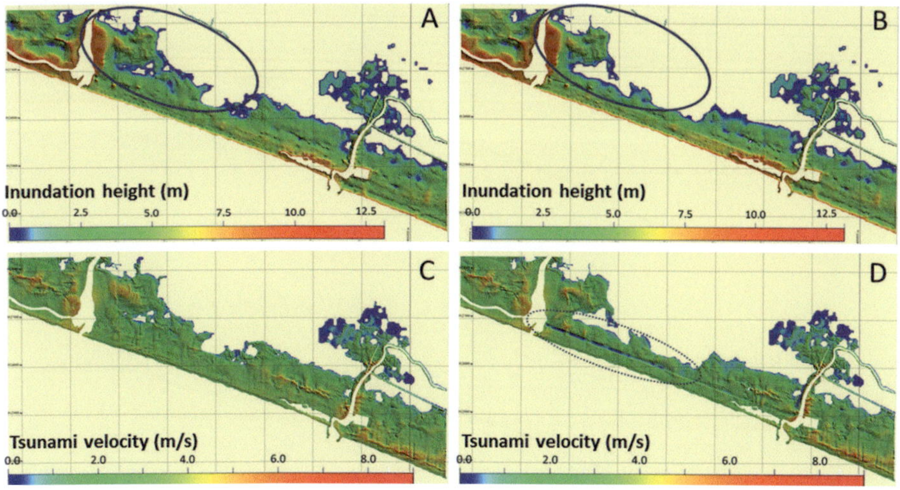

Fig. 3.9 The results of the numerical simulation for tsunami inundation and tsunami velocity for the case of no greenbelt (**a** and **c**) and with 50 m width greenbelt along the coast (**b** and **d**), which indicate that the existence of greenbelt reduces tsunami inundation area (see comparison between **a** and **b**) and reduces tsunami velocity significantly when it passes the greenbelt (inset area in **d**)

3.5.1 Effect of the Salty Vapor in the Plantation and Agriculture Near the Coast

The coastal forest planting sites, especially on the south coast of Java is an area less productive for agriculture. In Pacitan for instance, people living in the areas up to 500 m from the shoreline cannot use the area for agricultural purposes such as planting fruits or vegetables before the construction of the green belt. This is due to the salty marine aerosol prevailing in the wind during the daytime when the wind blows from the sea towards the land. The salty aerosol prevents fruits and vegetables seeds from growing as the soil quality are heavily altered when exposed (Fig. 3.10). Once the green belts are constructed, they act as natural barriers against wind-borne marine aerosol since it is filtered out by the tree branches bringing back potential growth in the affected areas.

3.5.2 Additional Economic Incomes

The existence of coastal forests enhances the tourism sector by not only providing shade and a place and area for solace for visitors but also it has triggered other economic benefits. A visit to the Pacitan District revealed that there has been a significant increase in the number of tourists visiting the beach when comparing the before and after the construction of the coastal pine tree forest. This yields an increase in income by 25 % in the communities after the construction of the greenbelt in general.

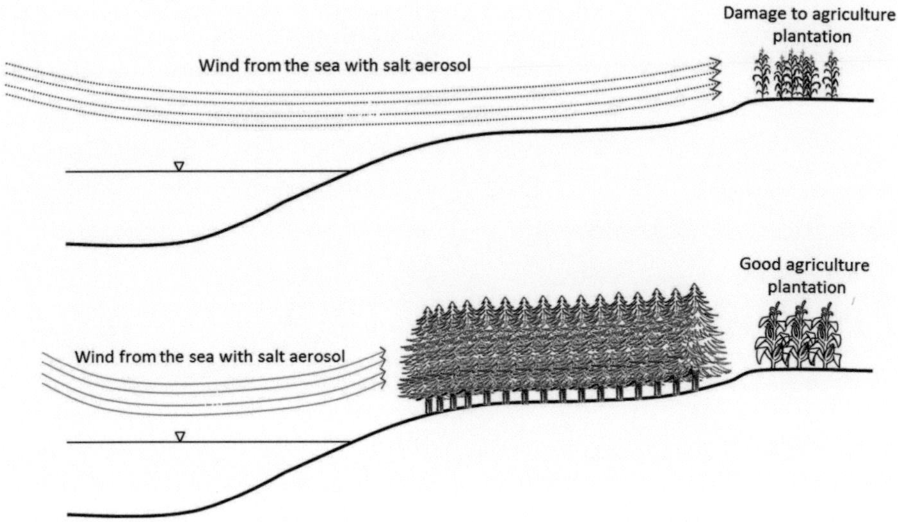

Fig. 3.10 An artist figure shows the protection of greenbelt to the plantation behind it from the wind of sea with marine aerosol

3.5.3 Replication in Other Areas

Replication of such a successful development of greenbelt in other areas in Indonesia needs encouragement not only from the government but also from the community itself by looking at successful cases. The authors started a field and comparative study program aimed at people in other areas following the successful experience in Pacitan, but it should not be forgotten the fact that aside of the indirect benefits to the communities, the original and main purpose to construct coastal forest is to mitigate the negative effects of future tsunamis hitting vulnerable areas along the Indonesian coastline.

3.6 Final Remarks

In this study field data has been used to generate the growth rate function as well as an input into the numerical model to assess tsunami hazard in vulnerable areas and where coastal forest construction can be undertaken. Moreover, we have presented a tsunami mitigation approach by using coastal forest as protective greenbelts in some areas in Indonesia. The application of this natural /ecosystem based approach will improve the quality of the coastal environment through the ecosystem service they provided to mitigate future tsunami impacts, while at the same time triggering a number of economic benefits to the local community. The replication of this approach can be highly recommended not only for implementation in Indonesia but also in other tsunami prone areas worldwide.

Acknowledgement The authors thank for the partial funding support from the Ministry of Marine Affairs and Fisheries of Indonesia.

References

BNPB-National Agency for Disaster Management (2011) Tsunami risk map Indonesia

Borrero JC, Sieh K, Chlieh M, Synolakis CE (2006) Tsunami inundation modeling for western Sumatra. Proc Natl Acad Sci PNAS 103(52):19673–19677

Borrero J, Weiss R, Okal EA, Hidayat R, Suranto DA, Titov T (2009) The tsunami of the 2007 September 12, Bengkulu Province, Sumatra, Indonesia: post-tsunami field survey and numerical modeling. Geophys J Int 178:180–194

Diposaptono S, Muhari A, Imamura F, Koshimura S, Yanagisawa H (2013) Impacts of the 2011 East Japan tsunami in the Papua region, Indonesia: field observation data and numerical analysis. Geophys J Int 194:1625–1639

Fritz HM, Kongko W, More A, McAdoo B, Goff J, Harbitz C, Uslu B, Kalligeris N, Suteja D, Kalsum K, Titov V, Gusman A, Latief H, Santoso E, Sujoko S, Djulkarnaen D, Sunendar H, Synolakis CE (2007) Extreme runup from the 17 July 2006 Java tsunami. Geophys Res Lett 34(12):5 pp. doi:10.1029/2007GL029404

Fujii Y, Satake K (2006) Source of the July 2006 West Java tsunami estimated from the tide gauge records. Geophys Res Lett 33(24). doi:10.1029/2006GL028049

Gayer G, Leschka S, Nhren I, Larsen O, Gnther H (2010) Tsunami inundation modelling based on detailed roughness maps of densely populated areas. Nat Hazards Earth Syst Sci 10:1679–1687. doi:10.5194/nhess-10-1679-2010

Gusman AR, Tanioka Y, Matsumoto H, Iwasaki S (2009) Analysis of the tsunami generated by the great 1977 Sumba earthquake that occurred in Indonesia. Bull Seismol Soc Am 99(4):2169–2179. doi:10.1785/0120080324

Hanifa NR, Sagiya T, Kimata F, Efendi J, Abidin HZ, Meilano I (2014) Interplate coupling model off the southwestern coast of Java, Indonesia, based on continuous GPS data in 2008–2010. Earth Planet Sci Lett 401:159–171

Harada K, Matsutomi H, Diposaptono S, Widagdo B (2011) Investigation of tsunami mitigation effect by coastal forest in Indonesia. In: Proceeding disaster management and climate change conference, Indonesia. Power Point/pdf in: http://www.soi.asia/data/event/20111027-disastermng/pdf/G3-2harada.pdf. Accessed Mar 2016.

Imamura F (1996) Review of tsunami simulation with a finite difference method. In: Yeh H, Liu P, Synolakis C (eds) Long wave run-up models. World Scientific, Singapore, pp 25–42

Imamura F (2009) Tsunami modeling: calculating inundation and hazard maps. In: Bernard EE, Robinson RA (eds) The sea, vol 15, Tsunamis. Harvard University Press, Cambridge, MA, pp 321–332

Imamura F, Muhari A, Mas E, Pradono MH, Post J, Sugimoto M (2012) Tsunami disaster mitigation by integrating comprehensive countermeasures in Padang City, Indonesia. J Disaster Res 7(1):48–64

IOC UNESCO -Intergovernmental Oceanographic Commission- (1997) Numerical method of tsunami simulation wit the leap-frog scheme. IUGG/IOC TIME project, manuals and guides, 35

Lovholt F, Kuhn D, Bungum H, Harbitz CB, Glimsdal S (2012) Historical tsunamis and present tsunami hazard in eastern Indonesia and the southern Philippines. J Geophys Res 117(B9):19 pp. doi:10.1029/2012JB009425

Lovholt F, Glimsdal S, Habitz C, Horspool N, Smebye H, de Bono D, Nadim F (2014) Global tsunami hazard and exposure due to large co-seismic slip. Int J Disaster Risk Reduct 10(Part B): 406–418. doi:10.1016/j.ijdrr.2014.04.003

Matsutomi H, Harada K, Widagdo AB, Diposaptono S (2011) Field test on the lodging and uprooting conditions of casuarinas and verification of those through the 2010 Mentawai earthquake tsunami. J JSCE, Ser. B2 (Coastal Eng) 67(2):301–305 (in Japanese)

Matsutomi H, Yamaguchi E, Naoe K, Harada K (2012) Damage to reinforced concrete buildings and coastal trees due to the 2011 off the Pacific coast of Tohoku earthquake tsunami. Coastal Eng Proc 33:1848–1860. doi:http://dx.doi.org/10.9753/icce.v33.management.51

Muhari A, Diposaptono S, Imamura F (2007) Toward an integrated tsunami disaster mitigation: lessons learned from previous tsunami event in Indonesia. J Nat Disaster Sci 29(1):13–19

Muhari A, Imamura F, Natawidjaja DH, Diposaptono S, Latief H, Post J, Ismail FA (2010) Tsunami mitigation efforts with pTA in West Sumatra Province, Indonesia. J Earthq Tsunami 4(4):341–368

Muhari A, Imamura F, Koshimura S, Post J (2011) Examination of three practical run-up models for assessing tsunami impact on highly populated areas. Nat Hazards Earth Syst Sci 11:3107–3123

Muhari A, Muck M, Diposaptono S, Spahn H (2012) Tsunami mitigation planning in Pacitan, Indonesia: a review of existing efforts and ways ahead. J Tsunami Soc Int 31(4):244–267

Muhari A, Imai K, Sugawara D, Imamura F (2014) A method to determine the level 1 and level 2 tsunami inundation areas for reconstruction in Eastern Japan and possible application in pre-disaster areas. In: Santiago-Fandino V et al. (eds) Post-tsunami hazard, advances in natural and technological hazards research, vol 44, Vicente, pp 133–157

Ohira W, Honda K, Harada K (2012) Reduction of tsunami inundation by coastal forests in Yogyakarta, Indonesia: a numerical study. Nat Hazards Earth Syst Sci 12:85–95

Okal EA, Synolakis CE (2008) Far-field tsunami hazard from mega-thrust earthquakes in the Indian Ocean. Geophys J Int 172(3):995–1015

Okal EA, Synolakis CE, Kalligeris N (2011) Tsunami simulations for regional sources in the south China and adjoining seas. Pure Appl Geophys 168(6):1153–1173

Pardede T (2014) National development planning for disaster risk reduction in Indonesia: reviewing the 10 years progress. In: Proceedings of international seminar on the 10 years commemoration of the 2004 Indian Ocean Tsunami

Satake K, Nishimura Y, Putra PS, Gusman AR, Sunendar H, Fujii Y, Tanioka Y, Latief H, Yulianto E (2013) Tsunami source of the 2010 Mentawai, Indonesia earthquake inferred from tsunami field survey and waveform modeling. Pure Appl Geophys 170:1567–1582

Shuto N (1987) The effectiveness and limit of tsunami control forests. Coast Eng Jpn 30:143–153

Spahn H, Hoppe M, Vidiarina HD, Usdianto B (2010) Experience from three years of local capacity development for tsunami early warning in Indonesia: challenges, lessons and the way ahead. Nat Hazards Earth Syst Sci 10:1411–1429

Strunz G, Post J, Zosseder K, Wegscheider S, Mück M, Riedlinger T, Mehl H, Dech S, Birkmann J, Gebert N, Harjono H, Anwar HZ, Sumaryono R, Khomarudin M, Muhari A (2011) Tsunami risk assessment in Indonesia. Nat Hazards Earth Syst Sci 11:67–82

Taubenbock H, Goseberg N, Setiadi N, Lammel G, Moder F, Oczipka M, Klupfel H, Wahl R, Schlurmann T, Strunz G, Birkmann J, Nagel K, Siegert F, Lehmann F, Dech S, Gress A, Klein R (2009) "Last-Mile" preparation for a potential disaster – interdisciplinary approach towards tsunami early warning and an evacuation information system for the coastal city of Padang, Indonesia. Nat Hazards Earth Syst Sci 9:1509–1528. doi:10.5194/nhess-9-1509-2009

Tsuji Y, Imamura F, Matsutomi H, Synolakis CE, Nanang PT, Harada JS, Han SS, Arai K, Cook B (2012) Field survey of the East Java earthquake and tsunami of June 3, 1994. Pure Appl Geophys 144(3):839–854

UNISDR (2013) From shared risk to shared value – the business case for disaster risk reduction, global assessment report on disaster risk reduction, United Nations Office for Disaster Risk Reduction (UNISDR), Geneva, Switzerland

Yanagisawa H, Koshimura S, Miyagi T, Imamura F (2010) Tsunami damage reduction performance of a mangrove forest in Banda Aceh, Indonesia inferred from field data and a numerical model. J Geophys Res 115(C6). doi:10.1029/2009JC005587

Chapter 4
Post-Tsunami Assessment of Coastal Vegetation, with the View to Protect Coastal Areas from Ocean Surges in Sri Lanka

L.P. Jayatissa, K.A.S. Kodikara, N.P. Dissanayaka, and B. Satyanarayana

Abstract The magnitude and impacts of tsunamis and ocean surges are unpredictable. Careful examination of these events and their consequences provide an insight towards natural protection against these hazards; therefore better preparedness and disaster risk reduction are crucial in countries at risk. This paper attempts to demonstrate some of the lessons from the December 26, 2004 tsunami which struck Sri Lanka, one of the most severely affected countries in the Indian Ocean region. We explored the effects of the tsunami on coastal vegetation in affected areas, and assessed their recovery after the event. We also explored the economic uses of the coastal plants, with the view that plant species with high resilience over tsunami, could be used to establish green barriers against tsunamis and ocean surges while at the same time being of economic value to local communities.

This study revealed that good mangroves were able to stand up against tsunami, helping to reduce the risk. However, mangrove (re-) forestation should be essentially limited to those areas that are physically and environmentally able to host them and the extent of such areas along the coastline of Sri Lanka is less than one third of the total coastline. This study also identified non-mangrove species, which could be the potential elements for green barriers along the non-mangrove areas of the coastline. It is noteworthy that all these species except cultivated plants were found to be constituents of the natural seashore vegetation and dune forming vegetation in the past; however, currently their individuals are sparsely distributed probably as a result of anthropogenic pressure. We believe that the direct economic uses of those species as well as the importance of green barriers may motivate dwellers to restore the coastal green barriers along the non-mangrove areas of the coastline.

Keywords Sri Lanka • Tsunami • Green barriers • Dune vegetation • Mangroves

L.P. Jayatissa (✉) • K.A.S. Kodikara
Department of Botany, University of Ruhuna, Matara, Sri Lanka
e-mail: lpj@bot.ruh.ac.lk

N.P. Dissanayaka • B. Satyanarayana
Institute of Oceanography, University Malaysia Terengganu (UMT), 21030 Kuala Terengganu, Malaysia

V. Santiago-Fandiño et al. (eds.), *Tsunamis and Earthquakes in Coastal Environments*, Coastal Research Library 14, DOI 10.1007/978-3-319-28528-3_4

4.1 Introduction

After the tsunami struck on countries around the Indian Ocean on December 26, 2004, the importance of natural coastal barriers, as protection against extreme oceanic influences was subjected to comprehensive deliberations (Clarke 2005; Liu et al. 2005; Overdorf and Unmacht 2005; Odling-Smee 2005). As one of these natural barriers, mangroves gained popularity in fora focusing on questions such as 'How were mangrove ecosystems affected during the tsunami?', 'Are all mangrove forests comparable in withstanding the tsunami?', 'Did mangroves serve as a natural protective barrier for lives and properties, and if so how?' A preliminary qualitative assessment of impacts of the tsunami on mangrove ecosystems, addressed those questions and emphasized mangrove forests' protective function (Alexander and Baird 2007; Chatenoux and Peduzzi 2007; Dahdouh–Guebas et al. 2005a; Mazda et al. 2007), although their protective potential has been undermined by direct and indirect human impacts (Dahdouh–Guebas et al. 2005b; Duke et al. 2007). In combination with other risk reduction measures such as sea walls and early warning systems, mangroves are often cheaper than solely conventional solutions (Feagin et al. 2010) and provide additional benefits like food, timber and carbon sequestration.

In-depth studies, particularly quantitative assessments, on impacts of the tsunami on coastal vegetation may give insights on how to achieve sustainable rehabilitation and reconstruction of affected areas as well as on how to build natural barrier against extreme events in the future. Creating effective natural barriers in the tropics to reduce risk from storms and tsunamis has traditionally relied considerably on the use of mangroves (Pearce 1996; Badola and Hussain 2005; Dahdouh-Guebas et al. 2005c; Williams 2005). However, mangroves are not the only candidate suitable for creating natural barriers (Adger et al. 2005; Barbier 2006).

Mangroves generally grow along coastlines of sheltered bays, shallow lagoons, and river mouths in tropical and subtropical areas and the occurrence and diversity of mangroves depends on climate, and the hydrography and physiography of an area (Duke 1992; Duke et al. 1998; Smith 1992). Obviously, mangroves do not normally grow along the entire coastline of a particular country (Spalding et al. 1997) which raises the question as what are the non-mangrove plants which can form potential natural barriers and to what extent non-mangrove solutions can potentially serve as suitable natural barrier of a coastline.

Not only do coastal green barriers provide direct protection form extreme oceanic events, they also provide more indirect long term protection of the coast against gradual erosion. Therefore the loss of these green barriers will have severe impacts on the vulnerability of the lives and properties of coastal communities (Dahdouh-Guebas et al. 2002).

Taking Sri Lanka, one of the countries most severely affected by the 2004 tsunami, as a case study, this paper presents the main results from post-tsunami field observations. The main objective of this study was to search for non-mangrove plants, which can be used as potential elements in restoring natural green barriers in the coastal belt including areas naturally not protected by mangroves, by rating how the non-mangrove plant species have withstood and recovered from the disaster.

4.2 Methodology

4.2.1 Possible Factors Affecting the Impact of the Tsunami

4.2.1.1 Mangrove and Non-mangrove Areas

Mangroves in Sri Lanka are concentrated in muddy to sandy substrates of intertidal areas of lagoons, river mouth, and sheltered bays or estuaries (Jayatissa et al. 2002a). Until today the extent of mangroves for the entire country has not been quantified accurately and, according to available 'estimates', mangroves cover of Sri Lanka is around 12,000 ha (IUCN 2011). One of the significant causes of mangrove forest loss during the last few decades has been the unsustainable shrimp farming, particularly in the northwestern coastal belt in Sri Lanka. As Senarath and Visvanathan (2001) reported, 64 % of the mangrove area around Puttalam lagoon, Sri Lanka, was lost during the period from 1981 to 1992.

Following categorization of the coastline and their quantification for Sri Lanka were adopted from Feagin et al. (2010) and used in this study (Further details on the methodology are given under 'supporting information' in the online version of Feagin et al. 2010);

Potential mangrove areas – the areas where there are mangroves now or there were mangroves in the past.

Mangrove coastline – coastline segment having a potential mangrove habitat behind it. (The total length of such segments for the whole country or for a particular administrative division, refers to the 'mangrove coastline of the country or the particular administrative division').

Non-mangrove coastline – the coastline other than the mangrove coastline.

Area potentially protected by mangroves – the landward area within the first 1 km belt of the mangrove coastline (the total of all such areas for the whole country or in an administrative division is hereafter referred as the '**Mangrove-protected area'**)

4.2.1.2 Vulnerable Areas Along the Coast

It is widely accepted that the presence of natural barriers can reduce the risk of tsunami and wind generated waves. However, the elevation of the coastal land from the mean sea level is a more prominent factor, which determine the vulnerability of a coastal land area to tsunami and wind-generated waves. Following terminology regarding vulnerable areas along the coast as well as their quantifications given in Feagin et al. (2010) were also adopted and used in this study:

Vulnerable area – the coastal land area lying under the 10 m contour line

Vulnerable coastline – Coastline segment having a vulnerable area at landward side (The total length of all these coastline segments for the whole country or in an administrative division is hereafter referred as 'vulnerable coastline' of the country or in the particular administrative division respectively).

Geographically protected coastline – the coastline other than the vulnerable coastline.

4.2.2 Field Observation on the Survival and Recovery of Common Trees in Tsunami-Affected Areas

The coastal stretch from Kalutara (west coast) to Batticaloa (east coast), which covers all the major climatic zones of the island (Fig. 4.1), was selected for the post-tsunami field observations. Fifteen field sites, distributed along the coastline

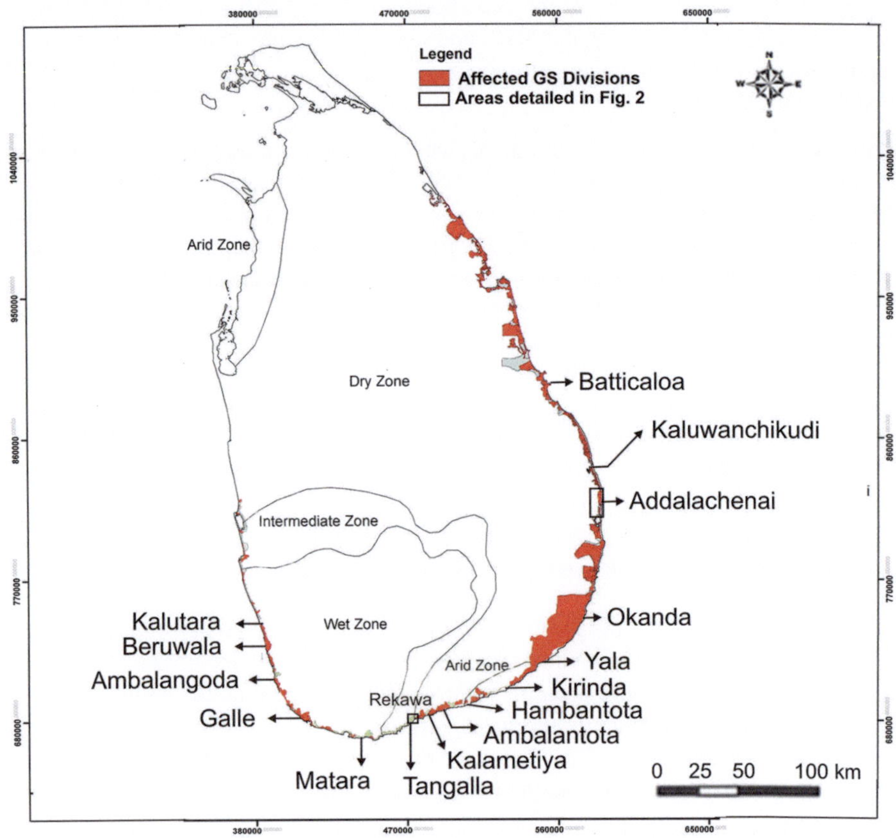

Fig. 4.1 Map of Sri Lanka showing tsunami affected GN divisions in eastern, southern and western provinces. Fifteen (15) study sites and locations of two areas detailed in Fig. 4.2. Major climatic zones are given according to Pemadasa (1996). Data on affected GN divisions obtained from the Department of Census and Statistics, Sri Lanka. http://www.statistics.gov.lk/SiteSearch.asp

representing all the major habitat types, were selected for thorough observations. At these sites the survival and recovery of plant species in non-mangrove areas affected by the tsunami was assessed. Each site was visited 14 days, 44 days and 134 days respectively after the tsunami. Both affected and non-affected plants species were recorded separately in all the major terrestrial habitat types including larger homesteads, agricultural lands, sand dunes, scrublands and forests. After 44 and 134 days, the recovery status of affected woody plants was also observed. A survey on the direct economic uses of different plant species recorded in affected areas was carried out.

4.3 Results

Feagin et al. (2010), sows that the total extent of the coastline of Sri Lanka is 1738 km. The breakdown of the entire Sri Lankan coastline into vulnerable and geographically protected parts and into mangrove and non-mangrove areas shows that the extent of the coastline with potential for mangrove vegetation in Sri Lanka is 538 km (30.9 %) and, out of that, 523 km falls within the vulnerable coastline. Furthermore, it shows that 1200 km of the total coastline its not suitable for mangroves and, out of that, 1057 km falls within the category of vulnerable coastline.

The map showing all GN divisions or GS divisions (‘*Grama Niladhari*’ division or ‘*Grama sevaka*’ divisions= the smallest administrative division) of Sri Lanka indicate that several GN divisions along the coastline were hardly, if at all, affected by the tsunami (Fig. 4.1). The overlay of this map with two GIS layers, one containing vulnerable areas and geographically protected areas, and the other containing mangrove and non-mangrove areas within a 1 km wide coastal belt, revealed that some of the non-affected GN divisions (where no causalities were reported after the tsunami), are not located on geographically protected areas but on areas protected by mangroves (see examples in Fig. 4.2). This is a direct confirmation of the protective function of mangroves.

As indicated in Dahdouh-Guebas et al. (2005c), this study also observed that the mangrove species were able to withstand the destructive forces of tsunami and therefore their survival against tsunami was subjected to much discussion. However, most of the terrestrial plants in coastal non-mangrove areas were unable to cope up with above three adversities and hence suffered physical and/or physiological damage. Physical damage varied from injuries to tender parts of the shoot system to uprooting of the whole plant whilst physiological damage varied from minor stress to the death of the plant. Although some species of woody plants survived the tsunami, their individuals were sparsely distributed (except those of Coconut –*Cocos sp.*- and *Casuarina sp.* plantations) among apparently dead plants and hence did not attract people’s attention. In the light of green barriers, these non-mangrove species will be more important than mangroves as they will be the only candidates for green barriers along more than two thirds of the coastline of Sri Lanka, although *Casuarina* species seem to have some problems as latter addressed in the Discussion.

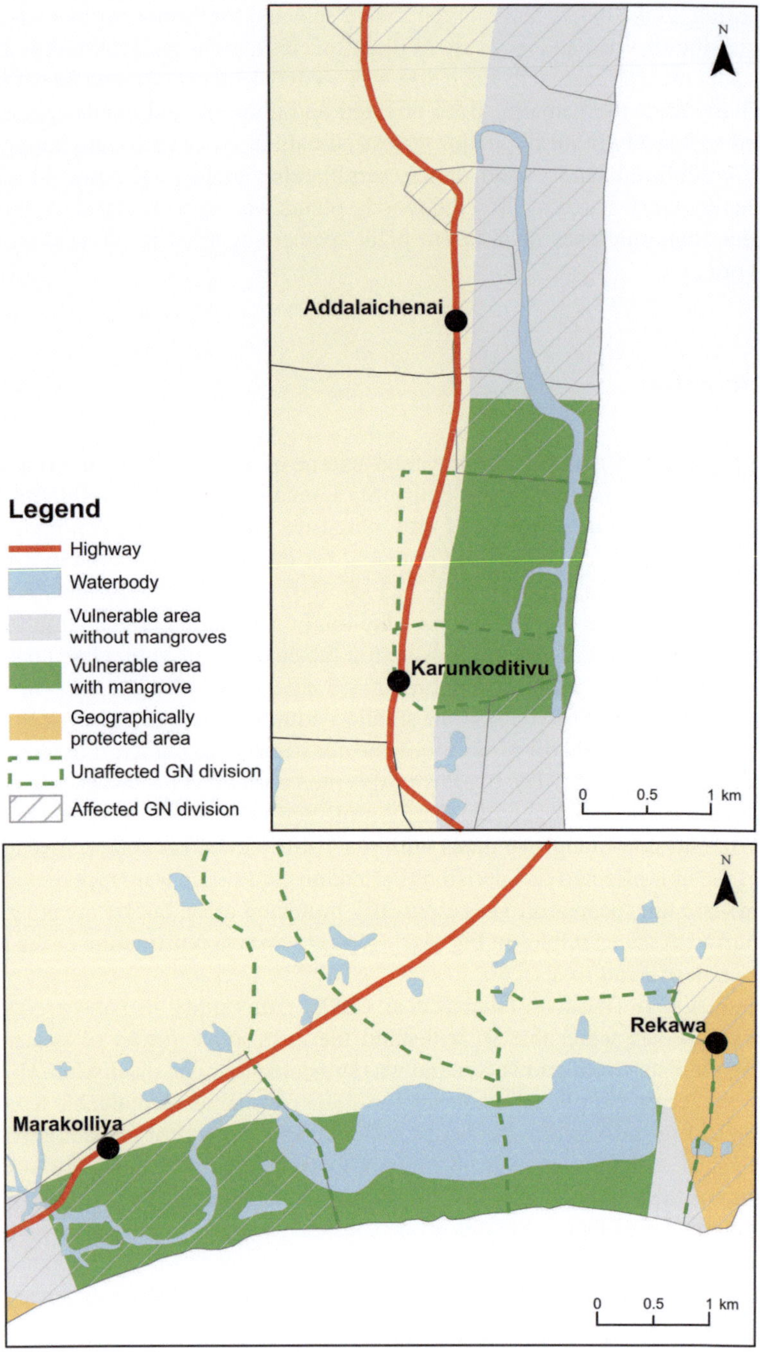

Fig. 4.2 Two examples of some GN divisions protected from the tsunami by mangroves whilst located in low-lying areas. Locations shown in Fig. 4.1 in *small rectangles*

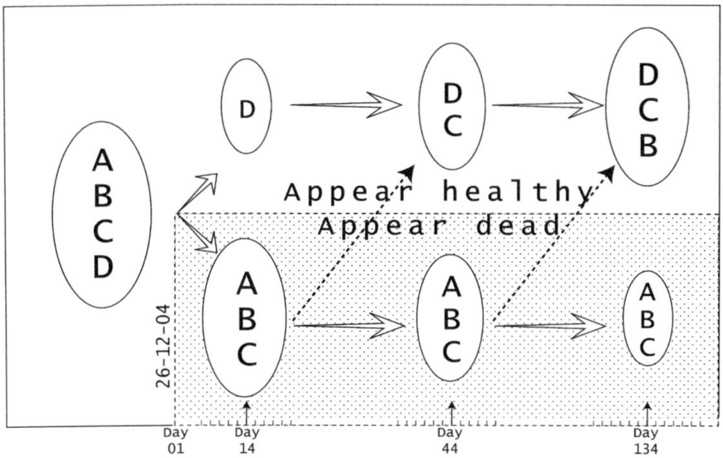

Fig. 4.3 Diagrammatic presentation of categorization (Group A–D) of some common plant species encountered at 15 study sites in tsunami-affected areas of Sri Lanka. Graphics based on the impacts and post-tsunami recovery after 14, 44, and 134 days. Group A: Species of plants affected and not recovered, Group B affected but some individuals gradually recovered after a month; Group C species which recovered in about 2 weeks, and Group D species which were not affected

During the field observations carried out 14 days after the tsunami, it was revealed that the coastal plants in tsunami-affected areas could be subdivided into two categories: 'apparently dead stands' and 'apparently healthy stands'. However, observations made 44 and 134 days after the disaster revealed that most of the plants in the former group had dropped all leaves, but were not dead. In fact, fresh buds and leaves were emerging gradually from the remaining shoot systems or from the base of the trunks. The time taken for this rejuvenation to occur, or the degree of damage was not similar amongst conspecific individuals, not even within the same site. For example, some individuals of the Palmyra palm *Borassus flabellifer* L., were dead in cases where the root system was exposed due to the removal of top soil by water currents, whilst individuals of which the root system was not exposed, remained healthy. Therefore, species of plants in affected areas were categorised as follows based on the impact of tsunami and recovery status (Fig. 4.3):

I. Group A. Species that were affected and that did not recover
II. Groups B and C. Species that were affected, and of which some individuals recovered gradually. The time taken for the recovery appeared to be dependent on the climatic and edaphic factors as well as on the extent of the damage, and ranged from more than a month (Group B) to less than 2 weeks (Group C);
III. Group D. Not-affected species.

Plants with intraspecific variation in their recovery status are thus grouped into category II; there was no intraspecific variation for plants belonging to category I or III. All the true mangrove species fall under category III. Jayatissa et al. (2002a) provides the full list of mangroves in Sri Lanka (not included in this Chapter). A list of common plants that belong to all four groups is given in Table 4.1 revealing that

Table 4.1 Vegetation from post-tsunami affected areas recorded during the present study (nomenclature mainly according to Mabberley 1987) and recovery status of Groups ***A*** to ***D***. Group ***A*** never recovered from the impact, Groups **B** and ***C*** recovered gradually (***B*** in more than a month while ***C*** in less than 2 weeks), and Group ***D*** not affected

Species	Common name (Sinhalese)	Type	Recovery status				Non-destructive economic uses
			A	B	C	D	
Wild plants							
1. *Acacia eburnea* (L.f.) Willd.	KatuAndara	T			●		Medicinal
2. *Asparagus racemosus* Willd.	Hatawariya	tW		●			Medicinal, + edible leaves
3. *Azadirachta indica* A. Juss.	Kohomba	T				●	Medicinal
4. *Azima tetracantha* Lam.	Wel Dehi	H				●	Medicinal
5. *Barringtonia asiatica* (L.) Kurz.	Mudilla	T				●	Medicinal
6. *Barringtonia racemosa* (L.) Spreng.	Diya Mudilla	T			●		Medicinal
7. *Bauhinia racemosa* Lam.	Maila	T				●	Medicinal
8. *Boerhavia diffusa* L.	Pita-sudu Pala	H			●		Medinal, + edible leaves
9. *Bulbostylis barbata* (Rottb.) Kunth ex Clarke	Uru hiri	H				●	Medicinal
10. *Caesalpinia bonduc* (L.) Roxb.	Kumburu wel	R			●		Medicinal
11. *Calophyllum inophyllum* L.	Domba	T				●	Medicinal
12. *Calotropis gigantea* (L.) R. Br.	Wara	S				●	Medicinal
13. *Canavalia rosea* (Sw.) DC.	Muduawara	R			●		Medicinal
14. *Carissa spinarum* L.	Heen-Karamba	S		●			Medicinal
15. *Cartunaregum spinosa* (Thunb.) Tirv. s. l	Kukuruman	sT		●			Medicinal
16. *Cassia auriculata* L.	Ranawara	S			●		Medicinal + edible leaves & flowers
17. *Cassine balae* Kosterm.	Neralu	T				●	Medicinal
18. *Cerbera manghas* L.	Gon Kadudru	T			●		Medicinal
19. *Clerodendron inerme* (L.) Gaertn.	WalGurenda	S				●	Medicinal
20. *Crateva* adansonii DC.	Lunuwarana	sT			●		Medicinal
21. *Cynodon dactylon* (L.) Pers.	Ruha	H				●	Medicinal
22. *Cyperus rotundus* L.	Kalanduru	H				●	Medicinal
23. *Dichrostachys cinerea* (L.) Wight & Arn.	Andara	sT			●		Medicinal
24. *Euphorbia antiquorum* L.	Daluk	S				●	Medicinal
25. *Exallage auricularia* (L.) Bremek.	Getakola	H			●		Medicinal
26. *Ficus benghalensis* L.	Maha Nuga	T			●		Medicinal

(continued)

Table 4.1 (continued)

Species	Common name (Sinhalese)	Type	Recovery status				Non-destructive economic uses
			A	B	C	D	
27. *Gmelina asiatica* L.	Demata	sT		●			Medicinal
28. *Gloriosa superba* L.	Niyangala	R			●		Medicinal
29. *Hedyotis auricularia* L.	Getakola	H			●		–
30. *Hernandia nymphaeifolia* (Presl.) Kubitzki	Palatu	sT			●		–
31. *Hibiscus tiliaceus* L.	Belipatta	S				●	Medicinal
32. *Hydrophylax maritima* L. f.	MuduGetakola	H				●	Medicinal
33. *Ipomoea pes-caprae* (L.) R. Br.	BimThamburu	H				●	Medicinal
34. *Justicia betonica* L.	SuduPuruk	H			●		Medicinal
35. *Launaea sarmentosa* (Willd.) Sch. Bip. ex Kuntze	–	H/R				●	–
36. *Macaranga peltata* (Roxb.) Müll. Arg.	Kenda	T	●				Medicinal
37. *Manilkara hexandra* (Roxb.) Dubard	Palu	T		●			Medicinal
38. *Maytenus emarginata* (Willd.) Ding Hou	Katu Pila	H			●		Medicinal
39. *Morinda citrifolia* L.	Ahu	sT			●		Medicinal
40. *Morinda coreia* Buch.- Ham.	Ahu	sT				●	Medicinal
41. *Opuntia dillenii*. (Ker-Gawl.) Haw.	Katu-Pathok	S				●	Medicinal
42. *Pandanus tectorius* Parkins. ex Z	Wetake	S				●	Medicinal
43. *Pedalium murex* L.	Eth-nerenchi	H				●	Medicinal
44. *Phoenix pusilla* Gaertn.	Indi / Walindi	T				●	Medicinal
45. *Phyla nodiflora* (L.) Greene	Hiramana Detta	H				●	Medicinal
46. *Pongamia pinnata* (L.) Pierre	Karanda	sT			●		Medicinal
47. *Premna latifolia* Roxb.	Maha midi	S				●	Medicinal
48. *Premna obtusifolia* R.Br.	Heen Midi	S				●	Medicinal
49. *Prosopis juliflora* (Sw.) DC.	–	T				●	–
50. *Pagiantha dichotoma (Roxb.) Markgraf*	Divi Kaduru	T			●		Medicinal
51. *Rungia repens* (L.) Nees	Sulu Nai	H				●	Medicinal
52. *Salvadora persica* L.	Malittan	T				●	Medicinal
53. *Scaevola plumieri* (L.) Vahl	Heen Takkada	H				●	–
54. *Scaevola taccada* (Gaertn.) Roxb.	Takkada	S				●	Medicinal
55. *Schleichera oleosa* (Lour.) Oken	Kon	T					Medicinal

(continued)

Table 4.1 (continued)

Species	Common name (Sinhalese)	Type	Recovery status A	B	C	D	Non-destructive economic uses
56. *Spinifex littoreus* (Burm. f.) Merr.	Maharawana Rewula	H				●	Medicinal
57. *Sansevieria zeylanica (L.) Willd.*	Niyanda	H			●		Medicinal
58. *Sarcostemma brunonianum* Wight & Arn. Ex Wight	Muvakiriya	C				●	Medicinal
59. *Syzygium cumini* Skeels	Madan	T			●		Medicinal + Edible fruits
60. *Tephrosia purpurea* (L.) Pers.	Gam-pila / Pila	H					Medicinal
61. *Terminalia catappa* L.	Kottamba	T				●	Medicinal + Edible seeds
62. *Thespesia populnea* (L.) Sol. ex Correa	Gan Suriya/ suriya	T				●	Medicinal
63. *Tylophora indica* (Burm. f.) Merr.	Mudu-Bin-Nuga	R				●	Medicinal
64. *Vigna trilobata* (L.) Verdc.	Bin me/ munwenna	R				●	–
65. *Ziziphus oenoplia* (L.) Miller	Heen Eraminia	S			●		Medicinal + Edible fruits
66. *Ziziphus rugosa* Lam.	Maha-Eraminia	S			●		Medicinal + Edible fruits
67. *Ziziphus mauritiana* Lam.	Masan	sT				●	Medicinal + Edible fruits
68. *Zoysia matrella* (L.) Merr.	–	H				●	–
Wild or cultivated plants							
69. *Anacardium occidentale* L.	Kadju/caju	T			●		Medicinal + edible fruits and seeds
70. *Borassus flabellifer* L.	Thal	T		●			Medicinal + edible parts
71. *Limonia acidissima* L.	Divul	T				●	Medicinal + edible fruits
72. *Moringa oleifera* Lam.	Murunga	T		●			Medicinal + edible fruits and leaves
73. *Pericopsis mooniana* (Thw.) Thw.	Nedun	T	●				Medicinal
74. *Tamarindus indica* L.	Siyambala	T			●		Medicinal + edible fruits
Cultivated plants							
75. *Artocarpus altilis* (Parkinson) Fosberg	Del	T		●			Medicinal + edible fruits
76. *Artocarpus heterophyllus* Lam.	Kos	T		●			Medicinal + edible fruits and seeds
77. *Casuarina equisetifolia* L.	Kasa	T				●	

(continued)

Table 4.1 (continued)

Species	Common name (Sinhalese)	Type	Recovery status				Non-destructive economic uses
			A	B	C	D	
78. *Citrus* L. spp.	Dehi /Dodam/ Naran	sT/S		●			Medicinal + edible fruits
79. *Cocos nucifera* L.	Pol	T				●	Medicinal + edible fruits
80. *Mangifera indica* L.	Amba	T			●		Medicinal = edible fruits
81. *Psidium guajava* L.	Pera	sT/S	●				Medicinal + edible fruits

H herbaceous species, *R* runner, *S* shrub, *sT* small tree, *T* tree, *tW* twining
Note: The list of impacts on mangroves is described by Dahdouh-Guebas et al. (2005c)

the number of woody species in group A is very low and these species are not common plants in dry or coastal areas, but in inland wet zone forests. Except in coconut or *Casuarina* plantations, healthy plants of group D were sparsely distributed among affected plants. It is evident by the photographs given as Plates 4.1 and 4.2.

About 90 % of the plant species that fall under group C or D, and therefore considered as potential elements for coastal green barriers, are of some economic value including as food source or used in indigenous medicine (Sugathadasa et al. 2008).

It was noted that a considerable area of the sandy beach or sand dunes along the south & east coasts of Sri Lanka were covered by *Casuarina* plantations. In almost all the plantations, there was no ground vegetation, but the soil was instead covered by accumulated *Casuarina* litter. Still, in few places alien invasive plants species such as *Lantana camara, Antigonon leptopus, Opuntia dillenii* dominated.

4.4 Discussion

The rise in number and intensity of many extreme natural ocean borne events is increasingly recognized as being the result of global and regional climate change (IPCC 2001). Unfortunately these events cannot be prevented nor always adequately predicted. Dahdouh–Guebas et al. (2005c) reported that the Indian Ocean area counted 63 tsunami events between 1750 and 2004 and more than three cyclones per year. Additionally, coastal erosion is an ongoing threat to Sri Lanka's coastline, particularly along the southern and western coasts (Dayananda 1992). Our best defence is to be adequately prepared and to try and reduce the risk and impact of these natural events. It is therefore crucial to learn from previous disasters to plan and take precautions, which can reduce their impact.

According to Feagin et al. (2010), only 8.9 % of the total coastline of Sri Lanka and 10.3 % of the total area of 1 km wide coastal belt is geographically protected against a potential tsunami of which the run up height is about 10 m. In other words,

Plate 4.1 (**a**) and (**b**). Palmyra and coconut palm trees located on the Potuvil beach which was exposed to the 2004 tsunami. Plate (**a**) and (**b**) show the same site respectively 14 and 134 days after the tsunami struck

Plate 4.2 (**a**) and (**b**). Surviving individuals of *Limonia acidissima* (**a**) and *Azadirachta indica* and *Cocos nucifera* (**b**) among apparently dead individuals of other species in tsunami affected areas at Tangalla study site

more than 90 % of the coastal area of Sri Lanka is vulnerable not only to tsunamis, but also to wind generated surges, coastal erosion by tidal wave actions and loss of land area due to rising sea level by global warming.

The shoreline of Sri Lanka is not a uniform entity and it covers a variety of tropical habitats such as coral reefs, seagrass and algal beds, seaweeds and mangroves, salt marshes, sand dunes and sea shore vegetation landward (Abeywickrema 1960; Pemadasa 1996). Many reports suggest that any of these habitats with the relevant communities developed up to the optimum level may function as barriers against extreme weather events and natural catastrophes (Fosberg 1971; Ewel et al. 1998; Dahdouh–Guebas et al. 2005c).

The suitability of mangroves as green barrier elements were emphasized in many articles based on superficial observations or qualitative & semi-quantitative studies on impact of tsunami (Willams 2005; Dahdouh-Guebas et al. 2005a, b, c; Danielsen et al. 2005). Several studies (Clarke 2005; Dahdouh-Guebas et al. 2005c; Gharibian 2015; Liu et al. 2005) and evidence given by mangrove dwellers from tsunami affected areas have confirmed that mangroves in many places of Sri Lanka have protected lives and properties against the 2004 tsunami. This was confirmed in the present study, since low-lying GN divisions located behind some of the good mangroves were not affected by the tsunami whilst adjoining GN divisions without protection by good quality or undisturbed mangroves were badly affected. This in turn can be justified as mangroves are known to display a large degree of adaptive capacity and being dynamic by nature (Reimers 1929; Chen and Twilley 1998; Dahdouh-Guebas et al. 2000; Duke 2001; Dahdouh-Guebas and Koedam 2002; Jayatissa et al. 2002b), a positive indication for their usefulness in restoring and protecting coastlines prone to mangrove growth.

Although the role of mangroves as a green barrier has been subjected to much discussions (Fosberg 1971; Ewel et al. 1998; Dahdouh–Guebas et al. 2005c; Liu et al. 2005; Gharibian 2015), the importance of the other categories of coastal vegetation as green barriers received less attention. On the other hand, more than two third of the total coastline of Sri Lanka is not suitable for mangroves. Therefore non-mangrove plant species with higher resilience that did survive the tsunami had to be distinguished through a proper in-depth analysis, as they could be the potential elements for green barriers along the non-mangrove areas of the coastline. This study contributes to fill that void.

Our results show that almost all individuals of a large number of non-mangrove plant species also, which were in tsunami-affected areas, stood up without much harm. It proves that all these species could be potential elements for green barriers along non-mangrove areas of the coastline, which is 68.9 % of the total coastline of Sri Lanka. If these species establish as a community with a proper structure on coastal areas, it could be a good barrier against tsunami and extreme weather events. Furthermore it could be a good source of medicinal plants used in indigenous medical system that is now not only used by local people but also by the tourist industry to give medical treatments for tourists who are coming to Sri Lanka solely to get indigenous medical treatments. However, these species have not been able to protect lives and properties against the recent tsunami, because individuals were distributed sparsely.

It is interesting to note that all the above mentioned species, except cultivated plants, were found to be constituents of the natural sea shore vegetation and dune forming vegetation (Abeywickrama 1960; Pemadasa 1996). Creeping plants in this tolerant group of species occupies the sandy area behind the direct impact of waves and tides, forming a thick carpet that accretes sand. This creeping carpet gradually leads to low shrubs and then to larger shrubs forming a dense zone where the soil is more stable; behind this shrub zone, on higher grounds there is a littoral woodland formed by tree components. These zones are physiognomically comparable but floristically different in the various climatic zones. In sand dune vegetation, the zone of creeping carpet is wider and at least some plants in other zones appear stunted. However, during the last few decades, these communities with a clear zonation were not there because, there was no coastal belt in Sri Lanka left for the natural coastal vegetation to be grown without disturbances. Wherever possible, coastal vegetation belts were encroached and used up to the last meter for settlements and agricultural activities destroying the natural vegetation and enhancing the coastal erosion. This has opened an easy avenue for the recent tsunami to rush into inland areas, but isolated individuals of the natural vegetation survived as they had adapted to the habitat.

Casuarina plantations have been established as green barriers along the south and east coasts of Sri Lanka by governmental and non-governmental organizations. However, negative impacts of coastal *Casuarina* plantations are now emerging. By reviewing the relevant literature, Awale and Phillott (2014) have shown that *Casuarina* plantations on sand dunes or sandy beaches enhance the coastal erosion and become an impediment for nesting turtles. Leelamani (2014) has shown that interstitial organic matter produced by *Casuarina* litter, make the soils in *Casuarina* plantations hydrophobic, resulting many negative impacts on soil biota. Due to all those facts, *Casuarina* cannot be recommended as an element for green barriers along the coastal belt and hence the need of other non-mangrove species becomes more important.

This study shows the possibility for such risk reduction measures based on lessons drawn from post-tsunami assessments conducted on the coastal vegetation of Sri Lanka. These findings could be used to increase the natural coastal defence against tsunamis, as well as against wind-generated surges, coastal erosion by tidal wave actions, and loss of land area due to rising sea level by global warming. As a first step towards improving disaster preparedness, high risk areas along a coastline should be identified and construction should be avoided. Effective warning systems can also be established and maintained. In the long term, the mechanisms of natural defence systems should be maximized by protection, restoration and reconstruction of natural barriers such as coral reefs, mangroves, other seashore vegetation and sand dunes. At the same time, highways, settlements and other public utilities should be shifted from high risk areas to safer sites. As an example the southern high way that was built after the tsunami, is running through more inland areas.

Natural coastal defenses already exist and within certain limitations they can be more economically established in many areas than constructing engineered defense solutions. Furthermore, they may provide additional benefits such as food and timber. If coastal characteristics allow for mangrove plantations growth they could be placed alongside certain types of infrastructure creating the "hybrid engineering" approach. Natural barriers can add protection support to sea walls and break waters (Dayananda 1992).

Based on the findings this study provides important information and lessons on the on the recovery potential and zonation of the addressed natural coastal vegetation in seashore and sand dune areas in Sri Lanka. Likewise, it addresses their capabilities and potential use to withstand and provide protection in events like tsunami and sea surges while also their potential economic value. The best way to pay our tribute to all who perished or suffered in the recent tsunami is to heed these powerful lessons it offered us.

Acknowledgments We thank Dr. Amara Satharasinghe (Department of Census and Statistics, Sri Lanka) for his cooperation in getting statistics used in this study. We acknowledge the support given by the Vlaamse Inter-universitaire Raad (VLIR), GREEN DYKE Project ZEIN2008PR347.

References

Abeywickrema BA (1960) The vegetation of the lowlands of Ceylon in relation to soil. In: Tropical soils and vegetation. Proceedings of the Abidjan symposium, Paris, UNESCO, pp 87–92

Adger WN, Hughes TP, Folke C, Carpenter SR, Rockstrom J (2005) Social–ecological resilience to coastal disasters. Science 309:1036–1039

Alexander MK, Baird AH (2007) Natural barriers to natural disasters. Bioscience 57(2):102–103. doi:10.1641/B570202

Awale D, Phillott AD (2014) A review of the adverse impacts of *Casuarina* spp on coastal ecosystems and sea turtle nesting beaches. Indian Ocean Turtle Newsl 19:15–19

Badola R, Hussain SA (2005) Valuing ecosystem functions: an empirical study on the storm protection function of Bhitarkanika mangrove ecosystem, India. Environ Conserv 32(1):85–92

Barbier EB (2006) Natural barriers to natural disasters: replanting mangroves after the tsunami. Front Ecol Environ 4(3):124–131

Chatenoux B, Peduzzi P (2007) Impacts from the 2004 Indian ocean tsunami: analysing the potential protecting role of environmental features. Nat Hazards 40:289–304. doi:10.1007/s11069-006-0015-9

Chen R, Twilley RR (1998) A gap dynamic model of mangrove forest development along gradients of soil salinity and nutrient resources. J Ecol 86:37–51

Clarke AC (2005) Rebuilding after tsunami: our key challenges. Daily Mail, 02. 01. 2005, London, UK

Dahdouh-Guebas F, Koedam N (2002) A synthesis of existent and potential mangrove vegetation structure dynamics from Kenyan, Sri Lankan and Mauritanian case-studies. Meded Zitt K Acad overzeese Wet/Bull Séanc Acad r Sci Outre-Mer 48(4):487–511

Dahdouh-Guebas F, Verheyden A, De Genst W, Hettiarachchi S, Koedam N (2000) Four decade vegetation dynamics in Sri Lankan mangroves as detected from sequential aerial photography: a case study in Galle. Bull Mar Sci 67:741–759

Dahdouh-Guebas F, Zetterström T, Rönnbäck P, Troell M, Wickramasinghe A, Koedam N (2002) Recent changes in land-use in Pambala-Chilaw Lagoon complex (Sri Lanka) investigated using remote sensing and GIS: conservation of mangroves vs. development of shrimp farming. Environ Dev Sustain 4(2):185–200

Dahdouh-Guebas F, Van Hiel E, Chan JCW, Jayatissa LP, Koedam N (2005a) Qualitative distinction of congeneric and introgressive mangrove species in mixed patchy forest assemblages using high spatial resolution remotely sensed imagery (IKONOS). Syst Biodivers 2(2):113–119

Dahdouh-Guebas F, Hettiarachchi S, Lo Seen D, Batelaan O, Sooriyarachchi S, Jayatissa LP, Koedam N (2005b) Transitions in ancient inland freshwater resource management in Sri Lanka affect biota and human populations in and around coastal lagoons. Curr Biol 15(6):579–586

Dahdouh-Guebas F, Jayatissa LP, Di Nitto D, Bosire JO, Lo Seen D, Koedam N (2005c) How effective were mangroves as a defence against the recent tsunami? Curr Biol 15(12):443–447
Danielsen F, Sorensen MK, Olwig MF, Selvam V, Parish F, Burgess ND, Hiraishi T, Karunagaran VM, Rasmussen MS, Hansen LB, Quarto A, Suryadiputra N (2005) The Asian tsunami: a protective role for coastal vegetation. Science 310:643
Dayananda HV (1992) Shoreline erosion in Sri Lanka's coastal areas. Coast Conservation Department, Colombo, 72pp
Duke NC (1992) Mangrove floristics and biogeography. In: Robertson AI, Alongi DM (eds) Tropical mangrove ecosystems. American Geophysical Union, Washington, DC, pp 63–100
Duke NC (2001) Gap creation and regenerative processes driving diversity and structure of mangrove ecosystems. Wetl Ecol Manag 9:257–269
Duke NC, Ball MC, Ellison JC (1998) Factors influencing biodiversity and distributional gradients in mangroves. Glob Ecol Biogeogr Lett 7:27–47
Duke NC, Meynecke JO, Dittmann S, Ellison AM, Anger K, Berger U, Cannicci S, Diele K, Ewel KC, Field CD, Koedam N, Lee SY, Marchand C, Nordhaus I, Dahdouh-Guebas F (2007) A world without mangroves? Science 317(5834):41–42. doi:10.1126/science.317.5834.41b
Ewel KC, Twilley RR, Ong JE (1998) Different kinds of mangrove forests provide different goods and services. Glob Ecol Biogeogr Lett 7:83–94
Feagin RA, Mukherjee N, Shanker K, Baird AH, Cinner J, Kerr AM, Koedam N, Sridhar A, Arthur R, Jayatissa LP, Danny Lo S, Menon M, Rodríguez S, Shamsuddoha M, Dahdouh-Guebas F (2010) Shelter from the storm? Use and misuse of bioshields for managing natural disasters on the coast. Conserv Lett 3:1–11
Fosberg FR (1971) Mangroves versus tidal waves. Biol Conserv 4:38–39
Gharibian A (2015) Mangroves: an asset to treasure. CNRS international magazine. Ed. Saman Musachchio. Quarterly no 36:38–39
Intergovernmental Panel on Climate Change (IPCC) (2001) Climate change 2001: the scientific basis climate change 1995. Cambridge University Press, New York
IUCN (2011) An appraisal of mangrove management in micro-tidal estuaries and lagoons in Sri Lanka. IUCN Sri Lanka Country Office, Colombo, viii + 116pp
Jayatissa LP, Dahdouh-Guebas F, Koedam N (2002a) A review of the floral composition and distribution of mangroves in Sri Lanka. Bot J Linn Soc 138:29–43
Jayatissa LP, Guero MC, Hettiarachchi S, Koedam N (2002b) Changes in vegetation cover and socio-economic transitions in a coastal lagoon (Kalametiya, Sri Lanka), as observed by teledetection and ground truthing, can be attributed to an upstream irrigation scheme. Environ Dev Sustain 4(2):167–183
Leelamanie DAL (2014) Relationship between water repellency and interstitial organic material of sieve fractions from dry zone dune sands under *Casuarina equisetifolia*. Proc Peradeniya Univ Int Res Sess Sri Lanka 18:423
Liu PLF, Lynett P, Fernando H, Jaffe BE, Fritz H, Higman B, Morton R, Goff J, Synolakis C (2005) Observations by the international tsunami survey team in Sri Lanka. Science 308:1595
Mazda Y, Wolanski E, Ridd PV (2007) The role of physical processes in mangrove environments: manual for the preservation and utilization of mangrove ecosystems. Terrapub, Tokyo, p 598
Odling-Smee L (2005) Dollars and sense. Nature 437:614–616
Overdorf J, Unmacht E (2005) Why save the forests? Fear of big waves is no reason to plant mangroves. Newsweek CXLV 7:49
Pearce F (1996) Living sea walls keep floods at bay. New Sci 150(2032):7
Pemadasa MA (1996) Green mantle of Sri Lanka. Tharangee Printers, Colombo, p 242
Reimers E (1929) Constantine *de Sa's maps and plans of Ceylon (1624–1628). With an introduction, transcripts, notes and translations.* A.C. Richards Acting Government Printer, Colombo, pp iv–vi, 51–62
Senarath U, Visvanathan C (2001) Environmental issues in brackish water shrimp aquaculture in Sri Lanka. Environ Manag 27:335–348
Smith TJ III (1992) Forest structure. In: Robertson AI, Alongi DM (eds) Tropical mangrove ecosystems. American Geophysical Union, Washington, DC, pp 251–292

Spalding M, Blasco F, Field C (1997) World mangrove atlas. The International Society for Mangrove Ecosystems, Okinawa, p 178

Sugathadasa KSS, Jeevanandara PM, Devanarayana A, Pushpakumara DKNG (2008) A checklist of medicinal plants of Sri Lanka. Bandaranaike Memorial Ayurvedic Research Institute, Navinna, Sri Lanka. ICRAF Sri Lanka Programme, p 278

Williams N (2005) Tsunami insight to mangrove value. Curr Biol 15(3):73

Chapter 5
Shoreline and Coastal Morphological Changes Induced by the 2004 Indian Ocean Tsunami in the Katchal Island, Andaman and Nicobar – A Study Using Archived Satellite Images

Ali P. Yunus, Jie Dou, Ram Avtar, and A.C. Narayana

Abstract The December 24, 2004 Sumatra earthquake and Tsunami had caused large damage to the coastal environment in the Indian Ocean countries. Continuous monitoring of shorelines are needed to understand the causes and consequences of recent changes and to assess the long term impact of tsunami waves. Assessment of the shoreline and coastal morphological changes due to tsunami in Katchal Island have been lacking due to obstacles in the field data acquisition owing to their remote location. As access to the ground information is limited, the only possibility is the monitoring of shorelines from multi-temporal satellite images.

In this study, we demonstrate the methods used in extracting shorelines and analyzing their changes using the data from Indian Remote Sensing (IRS) satellites, the EO1-ALI and Landsat images in a GIS environment. Eight satellite images acquired between 2004 and 2014 where used for the shoreline change and coastal morphology analysis in the Katchal Island. The results showed that the island experienced extensive erosion and significant loss in land area of about 20 km^2. Erosion has been more prevalent than accretion at an average linear regression rate of ~ −13 m/year between 2004 and 2010. Net shoreline movement of more than 4 km landward has been observed at the western coast of the island. Regions of high net shoreline movements were associated with bay-mouth areas, and regions linked with coastal inlets. This study demonstrates the strong potential of archived satellite images for detecting shoreline movements in far-off islands. The results will likewise help in understanding the response and recovery of shorelines in Indian Ocean regions after the 2004 tsunami.

A.P. Yunus (✉) • J. Dou
Graduate School of Frontier Science, The University of Tokyo,
5-1-5 Kashiwanoha, Kashiwa 277-8563, Japan
e-mail: yunusp@csis.u-tokyo.ac.jp

R. Avtar
Institute for the advance study of sustainability (UNU-IAS), United Nations University, Tokyo 150-8925, Japan

A.C. Narayana
Centre for Earth & Space Sciences, University of Hyderabad, Hyderabad 500-046, India

V. Santiago-Fandiño et al. (eds.), *Tsunamis and Earthquakes in Coastal Environments*, Coastal Research Library 14, DOI 10.1007/978-3-319-28528-3_5

Keywords 2004 tsunami • Erosion • Remote sensing • Recovery • Andaman and Nicobar

5.1 Introduction

The physical configuration of world's coastal shoreline is rapidly changing in recent years due to coastal erosion and accretion. In some cases these changes are driven by storm-surge waves, high tides, near-shore currents, while in others the changes are in response to tsunamis and landslides. Although the coastal zones only comprise less than 3 % of world's land area, small changes in the shoreline position largely affect the coastal ecosystem and habitats as they are home to many million people. Policy makers, climate change scientists, planning engineers and researchers require synoptic information regarding types and trends of shoreline movements. With the global warming induced sea-level rise advances, continuous monitoring of shorelines are therefore needed to understand the causes and consequences of recent changes in coastal environments and to assess the long term impact of natural events. Shoreline change analysis based on ortho-rectified satellite imagery has been popular since the Landsat mission, however the long term and continuous mapping of shorelines is uncommon. Global archives of multi-spectral satellite imagery are now readily available and have great potential with the availability of sensors high to moderate spatial resolution.

In the context of 2004 great Sumatra earthquake, the rupture along 1200 km segment of megathrust fault earthquake and following tsunami that lasted for several minutes caused widespread destruction of coastal communities in 11 countries in the Indian Ocean region (Indonesia, Sri Lanka, India, Thailand, Maldives, Somalia, Myanmar, Malaysia, Seychelles and others). Besides the causalities to life and properties, coastal morphology of these countries has been altered significantly by the devastating tsunami within a very short time. These include but not limited to coastal land uplift, subsidence, coral destruction, sediment erosion and transportation.

A number of studies have been reported in the recent past whereby these types of investigation have been carried out using field surveys (eg. Meltzner et al. 2006; Narayana et al. 2007; Paris et al. 2010). The imprints of the coastal erosion seems to be visible more than 500 m from the shore at many places. The geomorphological impact of tsunami is evidenced at Banda Aech, Sumatra in the form of beach erosion and destruction of sand dunes (Paris et al. 2009). An average of 3.75 % reduction in total land area is quantified for the islands of Maldives, suggesting future stability issues of mid-ocean reef islands (Kench et al. 2006, 2008). Meltzner et al. (2006) and Narayana (2011) in their study also reported land subsidence and erosion in the islands of Andaman and Nicobar region, India. Morton et al. (2008) studied the morphological changes at Yala, Sri Lanka and noted that the tsunami currents eroded the beaches and sand ridges resulting in irregular coastlines.

Remarkable bathymetric changes were observed in the nearshore zones of Kirinda Harbor, Sri Lanka, as a result of wave scouring and deposition (Goto et al. 2011). Transport of significant volume of inner shore sediments landwards during the tsunami at Kerala coast, India was reported in Narayana et al. (2007) studies. Deposition of sand sheets by the tsunami waves were observed about 420 m inland from the shore at Kalppakam, India (Srinivasalu et al. 2007). Post event studies on the Andaman Sea coast of Thailand noticed that tsunamigenic deposits were not fully preserved in the later period either due to human activities or by continued erosion (Szczuciński 2012). In certain regions of Yala and Banda Aech, widespread sediment deposition after the tsunami event were also visualized (eg. Morton et al. 2008; Wong 2009). Such deposition are indicative of responsive and recovery processes of the coastal morphologies.

By employing remote sensing imagery coupled with field investigations, tsunamigenic short-term shoreline response in Aech coast, Sumatra and Phangnga coast, Thailand were carried out by a few researchers. Choowong et al. (2009) by analyzing a series of satellite images along the Phang-nga coast has shown in their studies a complete recovery of the beach zones within 2 years after the tsunami. By an integrated approach using satellite and topographic datasets, Meilianda et al. (2010) briefed the accretion of sediments at the west coast and erosion in the north-west coast of Banda Aech just after 6 months of tsunami. Liew et al. (2010) observed development of new coasts that closely resembling the old one within weeks after the tsunami at the Aech coast Sumatra. Yunus and Narayana (2014) analyzed shoreline rate changes between the years 2004 and 2013 for the Andaman and Nicobar using multi-spectral remote sensing data, which is the only long term shoreline change analysis in the Indian Ocean region after the tsunami, but is limited to a small geographical area (Trinkat Island). Their studies indicated that Trinkat Island is continuously changing after the tsunami in such a way that 4.64 km^2 of the coastal area was eroded in the period between 2004 and 2013.

Currently an understanding of the shoreline and coastal morphological changes due to tsunami in Katchal Island, Nicobar group, is lacking because of constraints in the acquisition of field data. As the access to the ground information is limited, the possibility is extraction of shorelines from multi-temporal satellite images. The Indian Remote Sensing Satellite (IRS) mission – Linear Imaging Self Scanning Sensors (LISS-III and IV), Landsat Thematic Mapper (TM), Enhanced Thematic Mapper Plus (ETM +), and Earth Observing Advance Land Imager (EO-1 ALI) sensors have been in the space for a long time orbiting the earth continuously and acquired images systematically, thus particularly suitable for monitoring and assessing coastal environments. In this paper, we demonstrate the shoreline and coastal morphological changes induced by the great Sumatra earthquake to the Katchal Island using archived satellite data. Eight time period images between 2004 and 2014 are analyzed and presented.

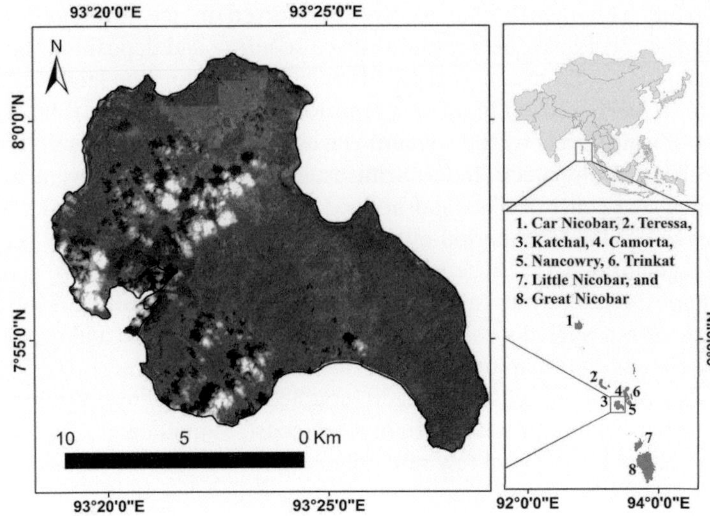

Fig. 5.1 Location and satellite image of Katchal Island within the Andaman and Nicobar group, India

5.2 The 2004 Tsunami and Katchal Island

On 26 December 2004, a 9.3 magnitude earthquake that occurred at 3.32° N latitude and 95.85° E longitude, at a depth of 30 m, generated a series of tsunami waves that swept across the Indian Ocean within few hours. The earthquake was occurred in the Sumatra–Andaman subduction zone where the Indo-Australian plate underthrusts the Burmese plate (Stein and Okal 2005). The sliding mechanism for the earthquake generation has been going for a long time, and as a result Burma plate had been stressed by subduction, and it rebounded from the frictional resistance on 26 December 2004 (Liew et al. 2010). The sliding caused a rupture more than 1200 km long and displacement of the seafloor above the rupture (Ammon et al. 2005). This earthquake was the fourth largest seismic event in the world since 1900. Waves 5–30 m high at the nearest Indian Ocean coasts were reported by the first *International Tsunami Survey Team* (Paris et al. 2007).

Katchal Island, separated from Camorta Island and Nancowry Island by Revello Channel is a part of Andaman and Nicobar group of islands located near the south-eastern coast of India (Fig. 5.1) The island has a population of 5386 people (NIC 2011), where most of the land area is covered with dense forests. The average annual rainfall is about 270 cm. Temperature of the island ranges between 25 °C and 30 °C with a relative humidity of 85 % (Dagar and Dagar 1991). Katchal Island that located near to the epicenter is one of the worst affected of the islands by the tsunamis. Tsunami waves reached the coasts of the Andaman and Nicobar islands 2–4 h after the earthquake. More than half the population of the Katchal Island was missing or killed during the event (Outlook 2005). And it also caused inundation of about 1787 ha of forest land resulting heavy ecological imbalance (Ramachandran et al. 2005).

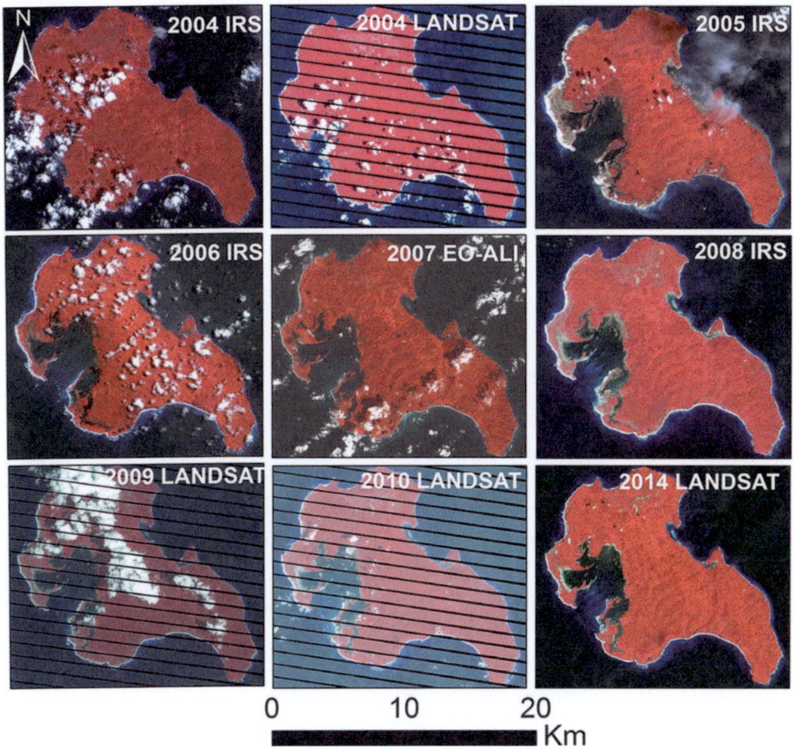

Fig. 5.2 Shoreline changes of Katchal Island; images taken prior and after the tsunami showing land erosion. Satellite images were obtained from NRSC and LP-DAAC

5.3 Methods

5.3.1 Satellite Images

Archived imageries are very useful in change detection studies. The readiness of images with repeated coverage of a single location made possible fast and inexpensive method of shoreline change statistics. In this research, we have used the optical satellite imageries between the years 2004 and 2014 obtained from the archives of National Remote Sensing Center (NRSC), India and the data distributed by the Land Processes Distributed Archive Active Center (LP DAAC), USGS/EROS (https://lpdaac.usgs.gov/) for analyzing the shoreline changes induced by tsunami to the Katchal Island (Fig. 5.2). Table 5.1 presents the data specification and acquisition date of the satellite images used. Because the data were obtained from different sources, all the images were re-projected to UTM coordinates and then geometric correction is performed. To perform this, pre-tsunami image of 2004 IRS P4 was used as a reference for image to image registration in the ENVI environment. The highest root mean square error (RMSE) recorded is 0.25 pixels. Achieving the RMS

Table 5.1 Data specification and land area changes to Katchal Island during 2004–2014

Image acquisition date	Sensor	Source	Total land area km²	Net change km²
2004/02/07	IRS LISS IV	NRSC	161.74	–
2004/03/30	LANDSAT ETM+	LP-DAAC		
2005/01/04	IRS LISS IV	NRSC	152.11	9.63
2006/04/29	IRS LISS IV	NRSC	145.42	16.32
2007/11/07	EO1- ALI	LP-DAAC	144.72	17.02
2007/02/07	IRS LISS III	NRSC		
2008/01/13	IRS LISS IV	LP-DAAC	144.72	17.02
2009/03/12	LANDSAT ETM+	LP-DAAC	144.41	17.33
2010/02/27	LANDSAT ETM+	LP-DAAC	144.35	17.39
2014/03/02	LANDSAT OLI	LP-DAAC	142.87	18.87

error values within the standard limits is necessary when using moderate resolution satellite images for change detection analysis (Chen et al. 2003). We resampled the 23.5 m resolution IRS data to 30 m resolution Landsat data using nearest neighborhood interpolation technique to obtain the consistency between the two data sets.

5.3.2 *Shoreline Extraction*

Automatic feature extraction algorithms were widely used for delineating shorelines from satellite images (eg. Mather 2004; Pardo-Pascual et al. 2012; Zhang et al. 2013). One of the most useful type of shoreline extraction method is edge enhancement (Lee and Jurkevich. 1990; Kumar et al. 2010). Two contrasting environments – water and land interface – allowed the edge enhancement algorithms to perform in superior manner in this situation. Several other researchers used supervised and unsupervised classifications to reliably extract the shoreline position (eg. Ekercin 2007; Ouma and Tateishi 2006). Manual digitization by analyzing proxies of shoreline, fuzzy logic, and band rationing techniques were also employed to precisely locate the shoreline positions (eg. Foody et al. 2005; Muslim et al. 2007; Sekovski et al. 2014; Yunus and Narayana 2014). Often, these techniques provide a realistic rapid shoreline representation; however a comparison of the techniques is beyond the scope of this paper.

Spatial filtering to the near infra-red band for enhancing the edges and then performed segmentation analysis to extract the shorelines have been undertaken. The overall workflow is described below: (i) image acquisition and preprocessing (see section satellite images), (ii) apply a non-linear edge enhancement 3×3 kernel Sobel filter that uses an approximation of Sobel function, (iii) segmentation of the images based on the Sobel enhanced image, and (iv) extraction of shoreline based on the boundary of segmented image. Wherever cloud obscures the data, we manually refined those regions with the previous year's shoreline position. Similarly the gaps in the 2009 and 2010 Landsat scenes due to scan line corrector (SLC) failure were corrected with the previous year data. An example of the shoreline extraction

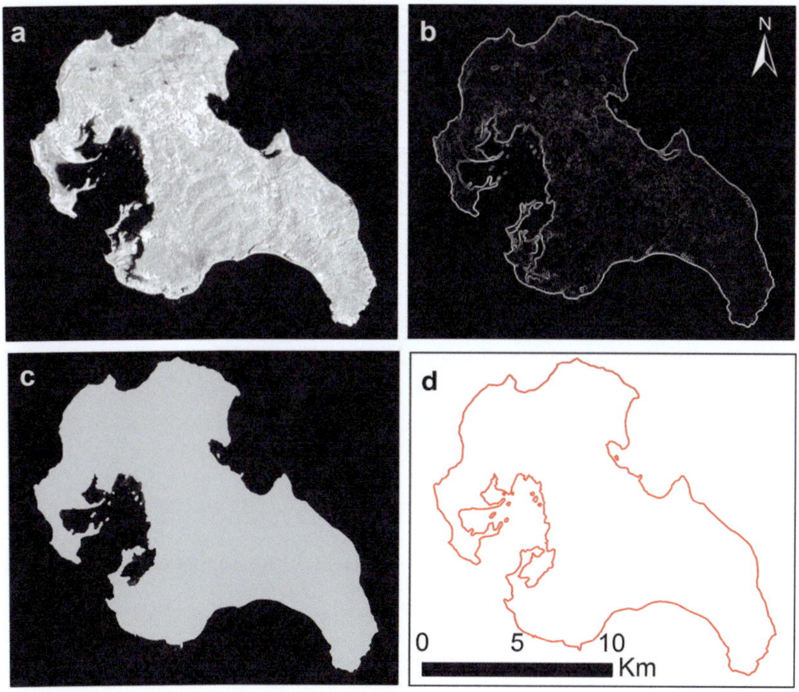

Fig. 5.3 Example of shoreline extraction method adapted in this study. (**a**) NIR-band image for Katchal Island, (**b**) after applying Sobel filter, (**c**) segmented imagery with 0 and 1 class, (**d**) shoreline boundary extraction based on the segmented image

in the study area is presented (Fig. 5.3). Tidal conditions at the satellite overpass time affects the accuracy of the shoreline rate of change estimation results. For accurate delineation of shorelines, previous researches took into account the position of high water lines (Kumar et al. 2010). The modeled estimate of highest high tide level for the region during the satellite overpass came up to be of 2.34 m (Yunus and Narayana 2014).

5.3.3 DSAS Shoreline Analysis

Digital shoreline Analysis System (DSAS) is developed as an add-on GIS software that computes change statistics from multiple historic shoreline positions (Thieler et al. 2009). One of the main advantages of using DSAS in shoreline change analysis is its ease of use, user friendly modules that can run within the ArcGIS platform. This tool generates transects perpendicular to the shorelines at a user specified distance. The transect intersection position of the shorelines along a baseline are then used to calculate the rate-of-change statistics. DSAS is capable of providing a variety of shoreline change statistics such as net shoreline movement (NSM), linear regression rate (LRR), and end point rate (EPR).

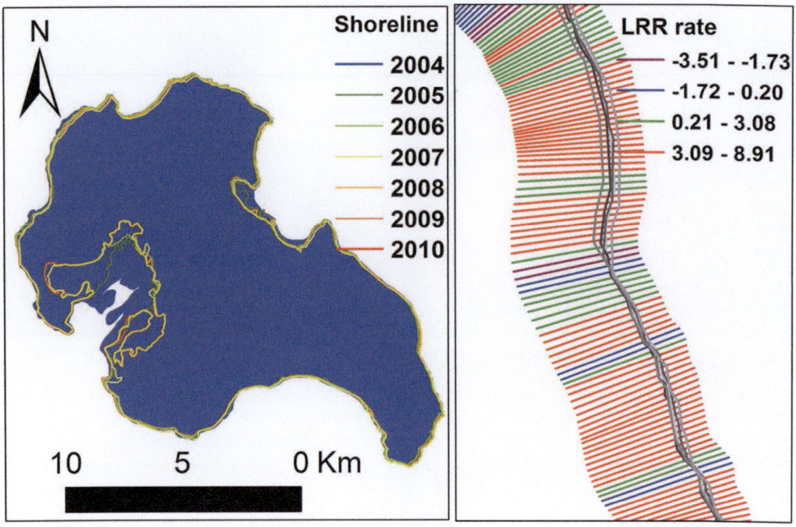

Fig. 5.4 Shoreline extracted for the years 2004–2010 and the location of transects (T) marked in the *left* figure. Example of DSAS based LRR rate in the *right* figure

The baseline in this study is established by buffering the 2004 shoreline position following the DSAS procedures (Himmelstoss 2009). A total of 1404 transects were generated for the study area by specifying a distance of 50 m. After careful consideration, 1328 transects were used in the shoreline change statistics. The NSM, that is the distance between oldest (2004) and youngest shoreline positions (2014), LRR the rate of shoreline change between 2004 and 2010, and the EPR were then estimated. Figure 5.4 shows the shoreline boundaries between 2004 and 2010 and the example of estimated LRR rate for a part of the study area.

5.4 Results and Discussion

Visual interpretation of the images captured between 2004 and 2014 suggests erosion in recent years (Fig. 5.2). Persistent land loss statistics in Katchal Island derived from the satellite images between 2004 and 2014 are shown in Table 5.1. The table shows that the island experienced substantial reduction in land area after the tsunami. The areal extent of the island before the tsunami was about 161.74 km^2 as measured from the images of February 2004. After the great earthquake and the following tsunami that affected the Katchal Island, the total land area decreased to 152.11 km^2; a change of about 10 km^2 as measured from images in January 2005. The measurement of total land areas from 2006, 2007, 2008, 2009, 2010 and 2014 satellite images is recorded as ~145 km^2, 144 km^2, 144 km^2, 144 km^2, 144 km^2, and 143 km^2 respectively. Overall, Katchal Island exhibited a significant diminution of land area of about 19 km^2 from 2004 to 2014, although partial recovery of straight

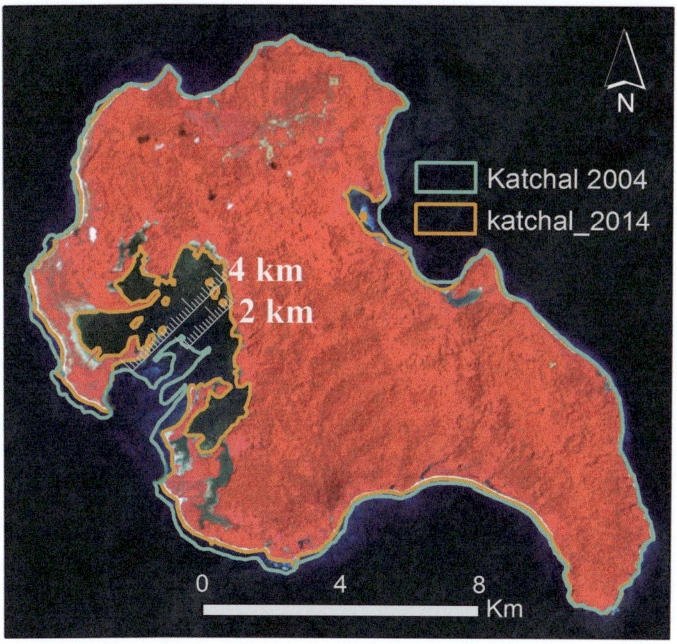

Fig. 5.5 Net shoreline movement (NSM) between 2004 and 2014 along the Katchal Island shown in *line colors*

beaches at some locations over a decade is noticed; but these changes were diffuse, with erosion continues in other areas.

Net shoreline movement was calculated for Katchal Island between the oldest and youngest shorelines using DSAS. The average NSM recorded for 1328 transects is – 154 m. At some locations, the estimated NSM over 2004–2014 shows more than 4 km landward shift in the shoreline positions (Fig. 5.5). The LRR results obtained using DSAS is shown in Fig. 5.6a. LRR values ranges between –726 m/year (indicating high erosion) and 25 m/year (low accretion). The average linear rate of erosion recorded between the years 2004 and 2010 is –13 m/year. The highest erosion is recorded between the transect 880 and 920 followed by transects between 148 and 160.

To understand the damages caused by the tsunamigenic and non-tsunamigenic events in the Katchal Island, EPR rate is calculated for two separate periods. The EPR values for 2004–2005 represent the direct influence of tsunami waves, whereas EPR values of 2005–2014 analysis shows normal wind-wave-surge dynamics. The results of EPR analysis is shown in Fig. 5.6b and c. The average EPR recorded for 2004–2005 is found to be –90 m/year. Maximum landward shift recorded by EPR for 2004–2005 is ~ –3663 m/year. The average EPR recorded for 2005–2014 is –6 m/year, whereas the maximum landward shift and seaward shift recorded are –420 m/year and 21 m/year respectively.

In some parts of the tsunami affected Indian Ocean coasts such as Aech, Trinkat and Phang-nga, partial to full recovery of beaches are reported (Choowong et al.

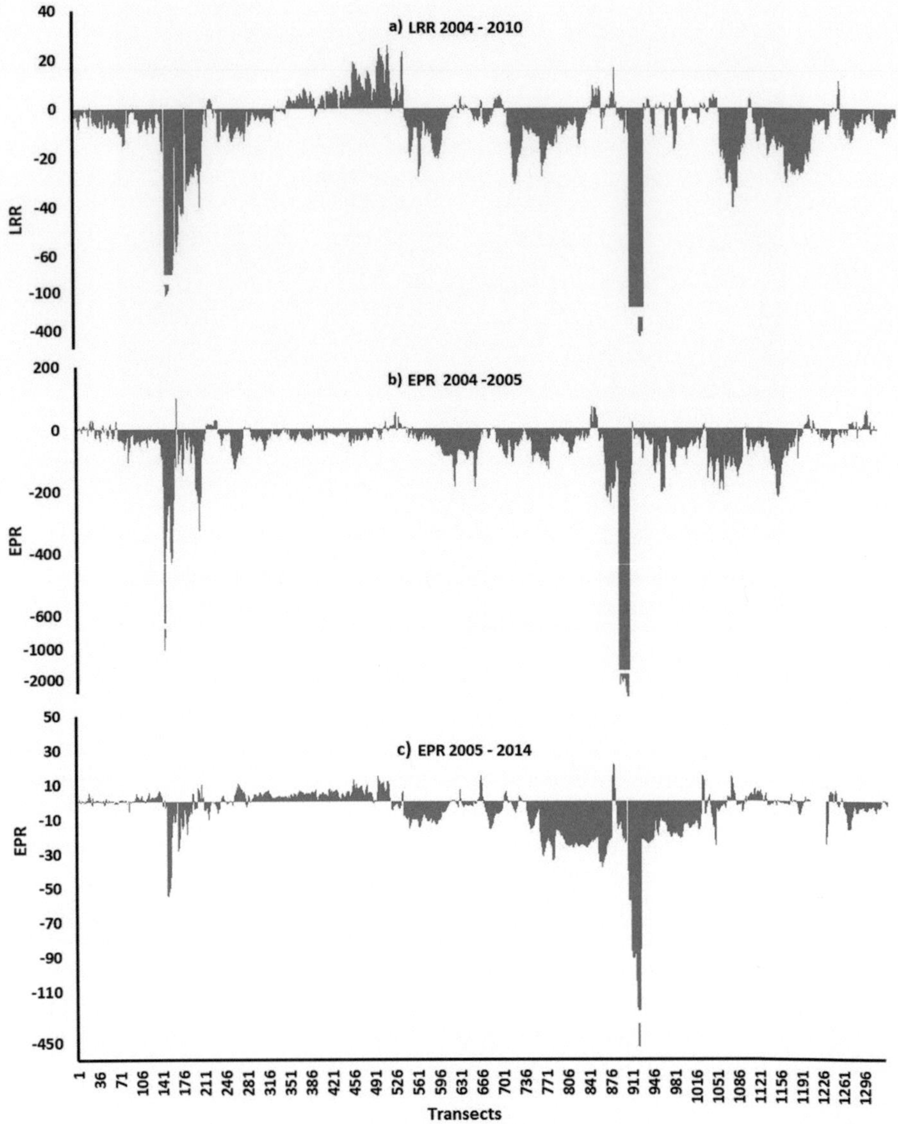

Fig. 5.6 (**a**), (**b**) and (**c**). Shore line rate change estimated for 1328 transects for Katchal Island. (**a**) Linear regression rate (LRR) for 2004–2010, (**b**) end point rate (EPR) between 2004 and 2005 and (**c**) EPR between 2005 and 2014

2009; Liew et al. 2010; Yunus and Narayana 2014). Although accretion is recorded at 538 transects in Katchal Island indicating a partial recovery, nowhere in the island the pace of recovery matches with the other regions. Morphological changes are clearly observed at two locations in the island: (i) the coastal bay opening in the western Katchal and (ii) the bay land in the eastern Katchal (Fig. 5.7). In the western

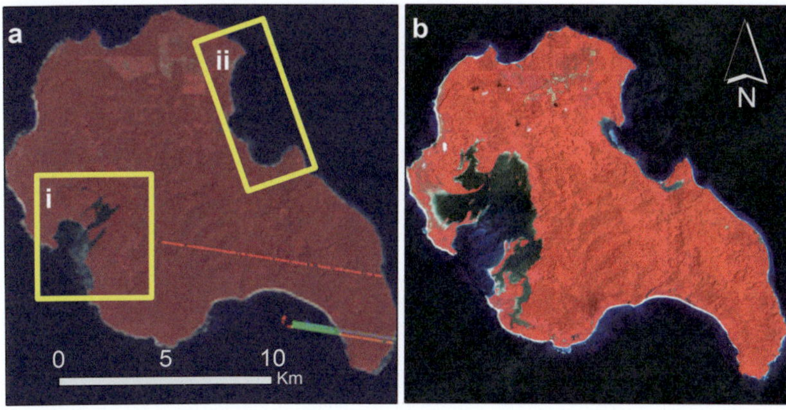

Fig. 5.7 Major morphological changes in the Katchal Island visualized at bay mouth lands: (**a**) pre-tsunami and (**b**) post-tsunami image

Katchal, the inlet at the mouth of the coast act as an opening for the tsunami waves to enter deep in to the land. Analysis of LRR and EPR shows that western part of the island is continuously eroding even after 10 years, although it is at slower rate. A possible reason may be the strong internal wave packets and tidal currents in the sea near Andaman during the monsoon periods (Apel et al. 1985).

We understand that the natural rate of shoreline movement at Katchal Island increased manifold after the tsunami. It is assumed that the removal of the vegetation cover, softening of the exposed bedrocks and unconsolidated materials induced by the tsunami causes the region vulnerable for erosion.

5.5 Concluding Remarks

The response and recovery of Katchal Island to the 2004 tsunami was not well understood earlier. Previous studies were constrained mainly to the ecological impact in the island (Ramachandran et al. 2005; Bahuguna et al. 2008). Our results based on the archived satellite images reveal that the island is vulnerable to erosion and unstable which may cause further erosion in future. Erosion rate is as high as −13 m/year, at some locations it reaches a maximum of −726 m/year. Evidence of accretion is noticed at some transects, which reflects recovery process, although it is at very slower rate. Major morphological changes are observed at the bay mouth areas.

A limitation of this study is arguably the usage of moderate resolution images for shoreline analysis. A minimum of 10 m (positioning error + greatest tidal effect) of movement is necessary to reliably detect the shoreline rates. As observed from the results, the average NSM was approximately -158 m, which is larger than the combined total of the RMSE and the highest tidal effect.

Considering the impact of tsunami to the Katchal Island and the observed shoreline movements of hundreds of meter, the methodology is thus still applicable. Our work demonstrates the strong potential of archived satellite images for detecting shoreline movements in far-off islands. We recommend continued monitoring and assessment of coastal features in the Katchal Island, since the region is highly dynamic in response to erosion.

References

Ammon CJ, Ji C, Thio HK, Robinson D, Ni S, Hjorleifsdottir V et al (2005) Rupture process of the 2004 Sumatra-Andaman earthquake. Science 308(5725):1133–1139

Apel J, Thomson D, Tilley D, Van Dyke P (1985) Hydrodynamics and radar signatures of internal solitons in the Andaman sea. John Hopkins APL Tech Dig 6(4):3330–3337

Bahuguna A, Nayak S, Roy D (2008) Impact of the tsunami and earthquake of 26th December 2004 on the vital coastal ecosystems of the Andaman and Nicobar islands assessed using RESOURCESAT AWiFS data. Int J Appl Earth Obs Geoinfo 10(2):229–237

Chen HM, Arora MK, Varshney PK (2003) Mutual information-based image registration for remote sensing data. Int J Remote Sens 24(18):3701–3706

Choowong M, Phantuwongraj S, Charoentitirat T, Chutakositkanon V, Yumuang S, Charusiri P (2009) Beach recovery after 2004 Indian ocean tsunami from Phang-Nga, Thailand. Geomorphology 104(3):134–142

Dagar HS, Dagar JC (1991) Plant folk medicines among the nicobarese of Katchal Island, India. Econ Bot 45(1):114–119

Ekercin S (2007) Coastline change assessment at the Aegean Sea coasts in Turkey using multitemporal Landsat imagery. J Coast Research 23(3):691–698

Foody GM, Muslim AM, Atkinson PM (2005) Super-resolution mapping of the waterline from remotely sensed data. Int J Remote Sens 26(24):5381–5392

Goto K, Takahashi J, Oie T, Imamura F (2011) Remarkable bathymetric change in the nearshore zone by the 2004 Indian Ocean tsunami: Kirinda Harbor, Sri Lanka. Geomorphology 127(1):107–116

Himmelstoss E (2009) DSAS 4.0 installation instructions and user guide. In: Thieler ER, Himmelstoss EA, Zichichi JL, and Ergul A (eds) Digital Shoreline Analysis System (DSAS) v4.0 An ArcGIS extension for calculating shoreline change: U.S. Geological Survey, Open-File Report: 2008-1278

Kench PS, McLean RF, Brander RW, Nichol SL, Smithers SG, Ford R et al (2006) Geological effects of tsunami on mid-ocean atoll islands: the Maldives before and after the Sumatran tsunami. Geology 34(3):177–180

Kench PS, Nichol SL, Smithers SG, McLean RF, Brander RW (2008) Tsunami as agents of geomorphic change in mid-ocean reef islands. Geomorphology 95(3):361–383

Kumar A, Narayana AC, Jayappa KS (2010) Shoreline changes and morphology of spits along southern Karnataka, west coast of India: a remote sensing and statistics-based approach. Geomorphology 120(3):133–152

Lee JS, Jurkevich I (1990) Coastline detection and tracing in SAR images. IEEE Trans Geosci Remote Sens 28(4):662–668

Liew SC, Gupta A, Wong PP, Kwoh LK (2010) Recovery from a large tsunami mapped over time: the Aceh coast, Sumatra. Geomorphology 114(4):520–529

Mather PM (2004) Computer processing of remotely-sensed images, 3rd edn. Wiley, Chichester

Meilianda E, Dohmen-Janssen C, Maathuis B, Hulscher S, Mulder J (2010) Short-term morphological responses and developments of Banda Aceh coast, Sumatra island, Indonesia after the tsunami on 26 December 2004. Mar Geol 275(1):96–109

Meltzner AJ, Sieh K, Abrams M, Agnew DC, Hudnut KW, Avouac J-P, Natawidjaja DH (2006) Uplift and subsidence associated with the great Aceh-Andaman earthquake of 2004. J Geophys Res 111:B02407. doi:10.1029/2005JB003891

M-Muslim A, Foody GM, Atkinson PM (2007) Shoreline mapping from coarse-spatial resolution remote sensing imagery of Seberang Takir, Malaysia. J Coast Res 23(6):1399–1408

Morton RA, Goff JR, Nichol SL (2008) Hydrodynamic implications of textural trends in sand deposits of the 2004 tsunami in Sri Lanka. Sediment Geol 207(1):56–64

Narayana A (2011) Tectonic geomorphology, tsunamis and environmental hazards: reference to Andaman-Nicobar islands. Nat Hazards 57(1):65–82

Narayana AC, Tatavarti Shinu N, Subeer A (2007) Tsunami of December 26, 2004 on the southwest coast of India: post-tsunami geomorphic and sediment characteristics. Mar Geol 242(1):155–168

NIC (2011) http://police.and.nic.in/katchal.htm. Accessed 10 Dec 2014

Ouma YO, Tateishi R (2006) A water index for rapid mapping of shoreline changes of five East African Rift Valley lakes: an empirical analysis using Landsat TM and ETM+ data. Int J Remote Sens 27(15):3153–3181

Outlook (2005) http://www.outlookindia.com/news/article/Over-6000--including-half-the-population-of-Katchal-missing/271300. Accessed 10 Dec 2014

Pardo-Pascual JE, Almonacid-Caballer J, Ruiz LA, Palomar-Vázquez J (2012) Automatic extraction of shorelines from landsat TM and ETM multi-temporal images with subpixel precision. Remote Sens Environ 123:1–11

Paris R, Lavigne F, Wassmer P, Sartohadi J (2007) Coastal sedimentation associated with the December 26, 2004 tsunami in Lhok nga, west Banda Aceh (Sumatra, Indonesia). Mar Geol 238(1):93–106

Paris R, Wassmer P, Sartohadi J, Lavigne F, Barthomeuf B, Desgages E et al (2009) Tsunamis as geomorphic crises: lessons from the December 26, 2004 tsunami in Lhok Nga, west Banda Aceh (Sumatra, Indonesia). Geomorphology 104(1):59–72

Paris R, Fournier J, Poizot E, Etienne S, Morin J, Lavigne F, Wassmer P (2010) Boulder and fine sediment transport and deposition by the 2004 tsunami in lhok nga (western Banda Aceh, Sumatra, Indonesia): a coupled offshore–onshore model. Mar Geol 268(1):43–54

Ramachandran S, Anitha S, Balamurugan V, Dharanirajan K, Vendhan KE, Divien MIP, Vel AS, Hussain IS, Udayaraj A (2005) Ecological impact of tsunami on Nicobar Islands (Camorta, Katchal, Nancowry and Trinkat). Curr Sci 89(1):195–200

Sekovski I, Stecchi F, Mancini F, Del Rio L (2014) Image classification methods applied to shoreline extraction on very high-resolution multispectral imagery. Int J Remote Sens 35(10):3556–3578

Srinivasalu S, Thangadurai N, Switzer AD, Mohan VR, Ayyamperumal T (2007) Erosion and sedimentation in Kalpakkam (N Tamil Nadu, India) from the 26th December 2004 tsunami. Mar Geol 240(1):65–75

Stein S, Okal EA (2005) Seismology: speed and size of the Sumatra earthquake. Nature 434(7033):581–582

Szczuciński W (2012) The post-depositional changes of the onshore 2004 tsunami deposits on the Andaman Sea coast of Thailand. Nat Hazards 60(1):115–133

Thieler E, Himmelstoss EA, Zichichi JL, Ergul A (2009) Digital shoreline analysis system (DSAS) version 3.0. An ArcGIS© extension for calculating shoreline change: U.S. Geological Survey, Open-File Report 2005–1304

Wong PP (2009) Impacts and recovery from a large tsunami: coasts of Aceh. Pol J Environ Stud 18:5–16

Yunus AP, Narayana AC (2014) Short-term morphological and shoreline changes at Trinkat island, Andaman and Nicobar, India, after the 2004 tsunami. Mar Geod 38(1):26–39

Zhang T, Yang X, Hu S, Su F (2013) Extraction of coastline in aquaculture coast from multispectral remote sensing images: object-based region growing integrating edge detection. Remote Sens 5(9):4470–4487

Chapter 6
Mud Volcanoes in an Active Fore-Arc Setting: A Case Study from the Makran Coastal Belt, SW Pakistan

Iftikhar Ahmed Abbasi, Din Mohammed Kakar, Mohammed Asif Khan, and Ahmed Sana

Abstract The Makran coastal belt is over 1000 km long stretching from Iran to east of Karachi in Pakistan. A major active subduction zone known as Makran subduction zone defined by the under thrusting of the Arabian plate beneath the Eurasian plate runs parallel to the coastline. The subduction zone is associated with a thick accretionary sedimentary wedge deposited in an active fore-arc basin containing very thick detrital sediments contributed by the accretion of the subducting plate since Late Eocene time. The sediments in the fore-arc basin are fine-grained usually of clay size fraction deposited in highly fluidized conditions trapping methane gas. The fluidized mud diapirically moves upward along weak zones as mud volcanoes due to high sedimentation rates and escaping gas pressure piercing through the overlying sediment layers. The development of mud volcanoes shows a close relationship between the sedimentation rates, gas escape from sediments and tectonic activity. Mud volcanoes are found in abundance both onshore and offshore of the Makran Coast. Most of the onshore mud volcanoes are associated with active fault zones and are believed to be triggered by tectonic activity. Located in the hanging wall of an active subduction zone, the region is seismically highly active and occur-

I.A. Abbasi (✉)
Department of Earth Science, College of Science, Sultan Qaboos University, Muscat, Sultanate of Oman
e-mail: iftikhar@squ.edu.om

D.M. Kakar
Department of Geology, Baluchistan University, Sariab Road, Quetta, Baluchistan, Pakistan
e-mail: dinkakar@yahoo.co.uk

M.A. Khan
Karakorum International University, Gilgit, Pakistan
e-mail: masifk9@gmail.com

A. Sana
Department of Civil Engineering, College of Engineering, Sultan Qaboos University, Muscat, Sultanate of Oman
e-mail: sana@squ.edu.om

V. Santiago-Fandiño et al. (eds.), *Tsunamis and Earthquakes in Coastal Environments*, Coastal Research Library 14, DOI 10.1007/978-3-319-28528-3_6

rence of major earthquakes exceeding magnitude 7 is a common phenomenon (e.g., Mw=7.7 2013 Awaran, Mw=8.2 1945 Makran). The seismic activity of this scale is likely the major triggerer for the emergence of new islands off the Makran coast in the Arabian Sea on regular basis in the past history. The latest of these emerged in 2013 immediately after the Mw=7.7 Awaran earthquake.

In this chapter we review major mud volcanoes formed along the Makran coastal region by describing their distribution, evolutionary history and mechanism of their formation. The chapter also describes the sudden appearance of islands off Makran coast since 1945 major earthquake and their importance in the geological history of the coastal regions.

Keywords Mud volcanoes • Mud islands • Makran • Accretionary wedge • Chandragup

6.1 Introduction

The Makran coastal belt is ideally suited for the study of fluidized sediments and their behavior in response to seismicity and gaseous activity, especially in an active fore-arc setting (Wiedicke et al. 2001; Delisle 2004; Calves et al. 2010; Kassi et al. 2014). High slip rates accompanied by high sedimentation rates (White and Louden 1982; DeMets et al. 1990) lead to formation of thick sedimentary accretionary wedge comprised of under-compacted and over-pressured fine-grained sediments, which are highly susceptible to deformation to form mud volcanoes (Platt et al. 1985; Calves et al. 2010). Such mud volcanoes are reported both from the offshore and onshore side of the Makran coast (Sondhi 1947; Snead 1964; Tabrez et al. 1999; Kassi et al. 2014).

The Makran accretionary prism (Fig. 6.1) was built due to low-angle (2–3°) subduction of the Arabian Plate beneath the Eurasian Plate (Fig. 6.2) at an average slip rate of 3 cm/year (DeMets et al. 1990). Scrapping and accretion of oceanic-floor sediments along the subducting Arabian Plate since Late Cretaceous-Early Paleocene time resulted in a sedimentary pile over 6 km thick (DeJong 1982; Leggett and Platt 1984; Minshull and White 1989; Gaedicke et al. 2002). The development of modern Makran accretionary wedge since the Late Miocene time lead to under-thrusting of the older wedge sediments and uplift of the accretionary complex (Platt et al. 1985). A number of studies dealt with both the onshore and offshore parts of the wedge to describe enormous thickness of the sediments, mechanics of fluidized features, deformation within and adjacent to the wedge, seismicity and subduction behavior (Harms et al. 1984; Platt et al. 1985; Byrne et al. 1992; Wiedicke et al. 2001; Kukowski et al. 2001; Schluter et al. 2002; Delisle 2004).

The onshore wedge comprising many kilometers thick east-west oriented Cretaceous to Recent age deformed sedimentary rocks extend four hundred kilometer inland from the subduction zone constituting Makran Ranges (Figs. 6.1 and 6.2).

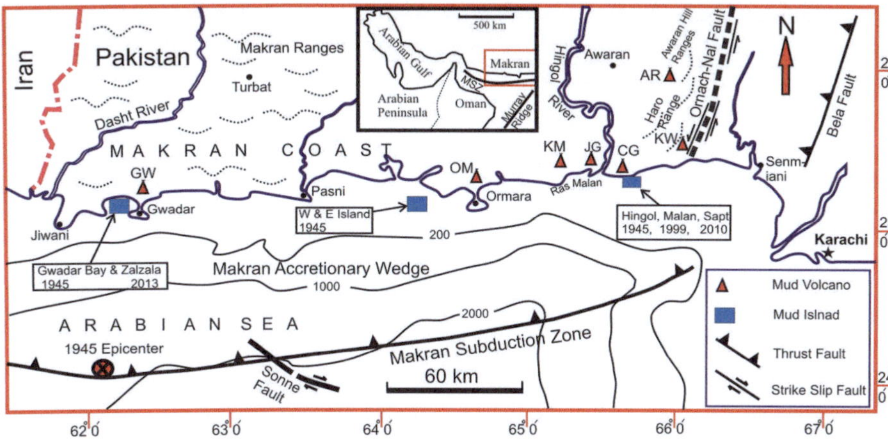

Fig. 6.1 Location map of Makran Coast showing distribution of mud volcanoes and temporary islands. Inset map shows regional features such as Muray Ridge and Makran Subduction zone. Mud volcanoes shown by triangles are AR-Awaran (25°52′09″N 65°45′25″E), KW-Kandewari (25°42′41″N 66°07′28″E), CG-Chandragup (25°27′06″N 65°52′26″E), JG-Jebel-u-Ghurab (25°23′58″N 65°44′53″E), KM-Kund Malir (25°25′55″N 65°26′29″E), OM-Ormara (25°21′34″N 64°37′10″E), GW-Gwadar (25°12′16″N 62°20′48″E). Mud islands are shown as rectangles, epicenter of 1945 Mw = 8.2 earthquake is shown by red circle (Modified from Sondhi 1947)

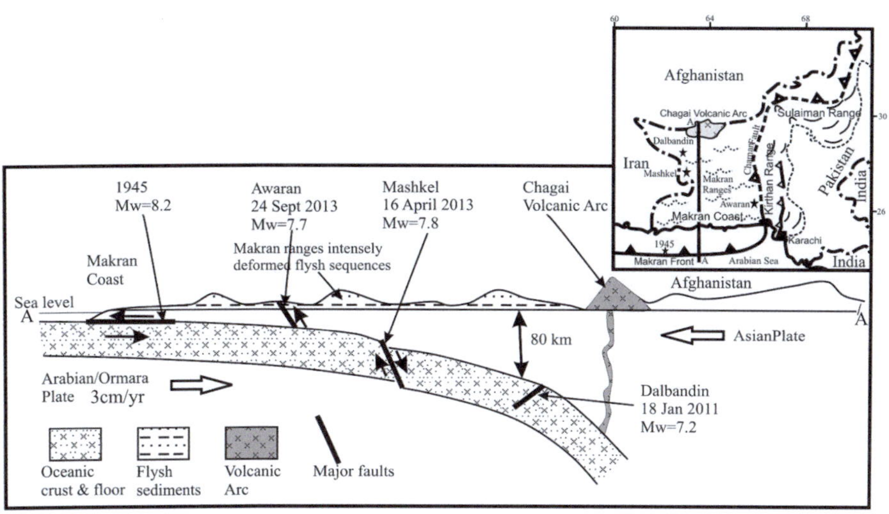

Fig. 6.2 Schematic north-south (A-A′) cross-section from Makran coastal front inland to Chagai Arc in the Asian Plate indicating association of major earthquakes with various tectonic lineaments. The inset map shows location of major tectonic and geographic features with major earthquakes Mw = 7 and above such as 1945 Mw = 8.2 and 2011 and 2013 shown as stars. Figure redrawn and modified after Roger Bilham personal Communication

Table 6.1 Generalized stratigraphy of the Makran Coast and adjacent areas (after Hunting Survey Corporation 1961; Kazmi and Abbasi 2008)

AGE		GROUP	FORMATION	LITHOLOGY
Quaternary	Holocene	MAKRAN GROUP	Jiwani Formation	Littoral deposits of shelly limestone, sandstone & conglomerate
	Pleistocene			Unconformity
			Ormara Formation	Soft & poorly consolidated mudstone & minor sandstone
				Unconformity
			Chitti Formation	Greenish grey, poorly bedded mudstone interbedded siltstone
Tertiary	Pliocene			Unconformity
			Talar/Hinglaj Formation	Sandstone, conglomerate, shale, shelly limestone
	Miocene	TURBAT GROUP	Parkini Formation	Greenish grey, poorly bedded mudstone interbedded thin sandstone
			Panjgur Formation	Grey sandstone & quartz arenite (basinal turbidites)
			Hoshab Shale	Bluish grey shale with thin sandstone intercalations
	Oligocene	Bathyal muds underlain by oceanic crust		

The older tightly folded and imbricated sediments are located inland of the Makran coast and are comprised of Cretaceous to Miocene mélange and turbidites, whereas relatively younger modern Makran wedge sequence contains very thick turbidite overlain by thick shelf sandstone (Table 6.1). Active subduction and thrusting resulted in frequent seismic activity and some of largest earthquakes have devastated the region causing tsunamis, land sliding and other surface disturbances (fracturing, fissuring, uplift and subsidence) that led to enormous losses of lives, properties and infrastructure in the historical past (Elliot 1857; Sondhi 1947; Bilham et al. 2007). One important consequence of the tectonic activity and seismicity in the Makran coastal region is formation of mud volcanoes (Sondhi 1947; Delisle et al. 2002; Delisle 2004; Bilham et al. 2007; Calves et al. 2010; Kassi et al. 2014). Of these, the ones offshore the Makran coast are particularly dramatic as they give rise to islands that suddenly emerge and then disappear over the time. The latest such mud island appeared about a km offshore the town of Gwadar immediately following the Mw=7.7 Awaran earthquake on September 24, 2013. Therefore the Makran accretionary wedge because of its enormous sediment thickness, high rate of sedimentation and tectonic deformation provides a unique opportunity to study diapiric flow of very fine-grain sediments in the form of mud volcanoes. In this chapter we review major mud volcanoes formed along the Makran coast by describing their distribution, evolutionary history and mechanism of their formation. The chapter will also describe the sudden appearance of islands off the Makran coast and their importance in the geological history of the coastal regions.

6.2 Geological Setting

The present day Makran accretionary wedge is 6–7 km thick highly deformed rock sequence at the prism front along the subduction zone. The prism is emergent about 100 km north of the subduction zone extending over 200 km northward onshore as east-west-trending sub-parallel ridges and valleys. The oldest part of the succession further inland contain Cretaceous to Paleocene mélange rocks of the Wakai mélange sequence tectonically overlain by a series of thrust sheets containing Paleocene and Eocene clastic and carbonate rocks (Hunting Survey Corporation 1961). An active volcanic arc lies about 400–600 km north of the prism known as the Chaghai Arc (Siddiqui 2004; Khan et al. 2010) (Fig. 6.2).

The coastal belt of Makran is comprised of rocks belonging to the Middle Miocene to Recent Turbat and Makran groups, divided into a number of formations (Harms et al. 1984; Kazmi and Abbasi 2008) (Table 6.1). The oldest rocks in the Makran coastal region are dark bluish grey abyssal shale named as the Hoshab Shale deposited during Oligocene-Middle Miocene time (Hunting Survey Corporation 1961; Harms et al. 1984). The Hoshab Shale is conformably overlain by over 2 km thick sandy, deep-sea fan complex and turbidites of the Panjgur Formation. The fan complex is covered by about 1 km thick, massive slope sediments of the Upper Miocene age mudstone of the Parkini Formation. From the Late Miocene to Early Pleistocene, concurrent with subduction related compressional structuring, there has been a period of slope, shelf and coastal plain progradation during deposition of sandstone-shale packages of Pliocene Talar/Hinglaj formations, and Plio-Pleistocene Chitti and Ormara formations (Hunting Survey Corporation 1961; Kazmi and Abbasi 2008).

The Makran accretionary wedge and offshore Oman Abyssal Plain is comprised of over 6–7 km thick pelagic and turbidite sequences which are divided on the basis of seismic reflection behavior into M1 (Makran-1) and M2 (Makran-2) sequences separated by a north dipping unconformity M (Makran) (Schluter et al. 2002). The mega sequence M1 is about 4.5 km thick and is comprised of hemipelagic and turbidite sediments of the Upper Cretaceous to Early Miocene in age possibly derived from the Eurasian and Afghan blocks. The mega sequence M1 corresponds to Hoshab Shale and Panjgur Formations exposed in Makran onshore belt. The overlying mega sequence M2 is estimated to be over 2 km thick and contains a mixed suite of turbidite accumulated due to growing Makran wedge since Middle Miocene time (Schluter et al. 2002). It corresponds to Late Miocene Parkini and younger formations in onshore Makran. Very high sedimentation rate during deposition of M2 of about 185 m/Myr (Schluter et al. 2002) and up to 2000 m/Myr during last 28,000 years (von Rad et al. 2000) resulted in initiation of liquidized mud movement as mud diapirs and mud volcanoes.

The Makran Coastal belt is seismically active due to its close association with a number of plate boundaries. Apart from being a major compressional regime defined by the subduction of the Arabian Plate below Eurasian Plate, another continental scale tectonic boundary named as Chamman/Ornach-Nal strike-slip fault between the Indian and Eurasian Plates is located along the eastern margin of the subduction zone. The east-west trending Makran subduction zone is cut by a NW-SE trending major left-lateral strike-slip fault, Sonne fault considered as a plate boundary between the Arabian Plate to the southwest and Ormara microplate in the northeast (Kukowski et al. 2000). The Little Murray Ridge in the Oman Abyssal Plain to the south of the Makran subduction zone defines a weak spreading zone (Fig. 6.1). Beside these plate boundaries, the accretionary wedge both onshore and offshore is highly deformed by a series of imbricate thrust slices (White and Louden 1982).

Historically very large earthquakes of $Mw = 7$ and above are reported from the Makran coastal belt and surrounding areas, located mostly along the plate boundaries. The largest instrumentally recorded earthquake was $Mw = 8.2$ on 28th November 1945 located along the Makran subduction zone offshore Makran coast (Figs. 6.1 and 6.2) (Sondhi 1947). The earthquake led to major tsunami affecting coastal belts of Pakistan/India as far as Mumbai and Oman. Coastal towns and Karachi port were affected and about 4000 people were killed. Since 2011 three major earthquakes larger than $Mw = 7$ have occurred in the area besides a number of moderate to low magnitude earthquakes. The seismicity of this magnitude leads to formation of onshore mud volcanoes and appearance of temporary islands in offshore Makran coast such as the ones associated with 1945, 1999, 2010 and 2013 earthquakes.

6.3 Mud Volcanoes Along Makran Coast

High sedimentation rates, common seismicity especially high magnitude earthquakes and presence of methane hydrates in deep marine sediments make Makran coastal belt both onshore and offshore an ideal place for formation of mud volcanoes. Mud volcanoes are reported from several places along the Makran coastal belt onshore, the most famous being the Chandragup mud volcanoes (Schluter et al. 2002; Delisle 2004; Kassi et al. 2014). A number of mud volcanoes are also reported from the offshore Makran coast and southern Indus Fan (Collier and White 1990; Delisle 2004; Calves et al. 2010).

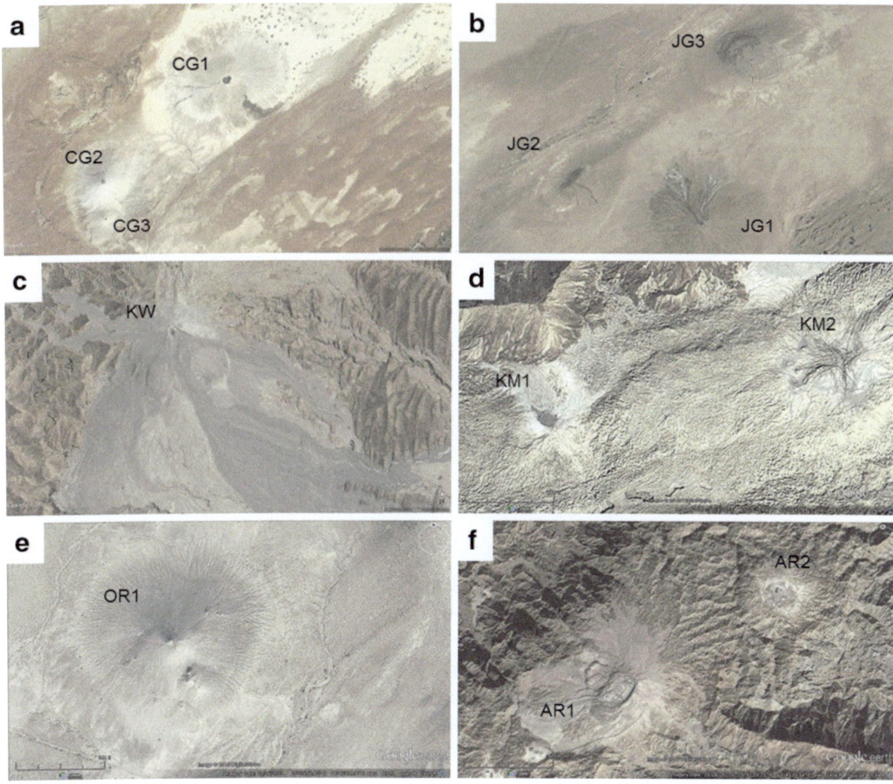

Fig. 6.3 Google images of various mud volcanoes on Makran Coast, (**a**) Chandragup mud volcano cluster showing Chandragup1,2,3 (CG1, CG2, CG3), (**b**) Jebel-u-Ghurab mud volcano cluster showing Jebel-u-Ghurab1,2,3 (JG1, JG2, JG3), (**c**) Kandewari mud volcano (KW1) is the largest mud volcano in the Kandewari cluster comprising six individual mud volcanoes, (**d**) Kund Malir cluster Km1 & Km2, (**e**) Ormara mud volcano containing a number of vents, (**f**) Awaran cluster of mud volcanoes AR1 & AR2 having a number of vents

6.3.1 Onshore Mud Volcanoes

Mud volcanoes along the Makran coast constitute most spectacular circular features at a number of localities especially along its eastern side such as near Ormara and Hingol (Figs. 6.1, 6.3a and 6.4a, b). The Chandragup mud volcanoes near Hingol are the largest one in Makran region (Figs. 6.3a and 6.4a). The cluster of three mud volcanoes named as Chandragup1, Chandragup2 and Chandragup3, each containing a number of subsidiary mud eruption vents, is located about 3.5 km north of the coast. Chandragup1 crater is 2.5 m deep (below crater top) having a water lake with gas bubbling in the center of mud volcano (Fig. 6.4c). The mud extrusion is episodic, however gas bubbles up regularly in water with low frequency. The crater diameter is about 30 m and the area of the water lake is 70×50 (1.09 km^2) meters. The crater is defined by circular walls which are partly eroded and collapsed

Fig. 6.4 Field photographs of Chandragup mud volcanoes, (**a**) Chandragup1 cone is the largest of all the mud volcanoes in the Makran coast, (**b**) Dead vent of Chandragup3, (**c**) Crater lake in Chandragup1 mud volcano (person in circle for scale), (**d**) mud bubbles-up due to methane escape through mud

(Fig. 6.4b). The total area of the volcano is about 1000×700 m (2.2 km^2) across. This appears to be the largest and oldest of the Chandragup mud volcano cluster. As mud spewing has been stopped for the last several years, there is no deposition of mud in its crater and the process of erosion has removed most of its crater deposits. Chandragup2 is the highest and the most prominent volcano in Makran region located about 4 km from sea coast (Fig. 6.3a). Its crater diameter is about 20 m, filled with viscous and sticky mud slurry. This is an active volcano extruding mud water slurry along with gaseous bubbles from its vent (Fig. 6.4d). The mud bubbles up and burst violently due to emanating gases since last many decades. The height and diameter of Chandragup2 has been increasing through time, as mud eruption continues and is deposited in its crater. Chandragup3 is 30 m above the average ground level located in close vicinity of Chandragup2. It is extinct and dried-up having a crater of about 14 m in diameter (Figs. 6.3a and 6.4b). Kassi et al. (2014) reported additional 11 subsidiary mud volcanoes associated with above described three major ones.

Apart from impressive Chandragup cluster, a number of other mud volcano fields namely Jebel-u-Ghurab, Kandewari, Kund Malir, Gwadar, Ormara and Awaran clusters are identified and described (Snead 1964; Reimann 1989; Delisle et al. 2002; Kassi et al. 2014) (Figs. 6.1 and 6.3b–f). All of these fields are comprised of two or more isolated conical mud volcanoes oriented along regional tectonic lineaments such as thrust faults (Fig. 6.1). The Jebel-u-Ghurab volcanic field is spread over an area of less than one square kilometer comprising four mud mounds which are one to eight meters high. The occurrence of the mud volcanoes in the core of an anticlinal structure suggests residual gas concentration in the crest of the anticline (Delisle et al. 2002). Mud water and gas intrusion is facilitated further by fold axis parallel fractures which are common in the area. Most of the volcanoes are dried-up and their craters are filled with water. The Kandewari field comprising six mud volcanoes is located in Haro Range about 30 km ENE of the Chandragup cluster (Figs. 6.1 and 6.3c). The mud volcanoes are spread over 20 km long NNE-SSW trending belt parallel to the Haro range in which four clusters are identified. The mud volcanoes are active emitting fluidized mud periodically. Mud volcanoes near the town of Gwadar are small size cones, the largest crater having a maximum diameter of 7 m. The volcanic cones are considered dormant as no fluidized activity is reported in the recent past. A cluster of active mud volcanoes near Kund Malir is reported for the first time by Kassi et al. (2014) (Figs. 6.1 and 6.3d). A total of 16 isolated cones spread over 5 km long NE-SW trending belt are reported from Malan Range. The largest crater in the field is measured as 25 m. The mud extruded from various cones coalesces to form thick mud deposits. Active volcanic cones are also located about 9 km NNE of Ormara coastal town named as Ormara mud volcanoes (Figs. 6.1 and 6.3e) (Delisle et al. 2002; Kassi et al. 2014). A total of eleven mud volcanoes are identified in three clusters spread over 9 km long belt trending E-W. The clusters are made up of a number of small cones which in certain cases combine together to form relatively large fields with a number of craters. North of Kandewari area in Awaran Hill ranges, at least eight active mud volcanoes are identified (Figs. 6.1 and 6.3f). The mud volcanoes are spread along a four kilometer long NNE-SSW oriented belt. Kassi et al. (2014) reported over 20 mud volcanoes in Awaran ranges within a 58 km long N-S oriented belt parallel to the strike of the Awaran Hills. The crater diameter of these mud volcanoes ranges from less than one meter to over eighty five meters.

This brief overview suggests that the mud volcanoes are not restricted only to the coastal belt of the Makran region but spread over a large area at least over 70 km inland (Fig. 6.1). Almost all of the above mentioned mud volcanoes particularly the large ones such as Chandragup, Kandewari and Awaran mud volcanoes are located along tectonic lineaments parallel to the mountain range trend.

6.3.2 Offshore Mud Volcanoes

In the offshore Makran, mud mounds formed due to fluidized processes related to rapid sediment deposition and active tectonics in the area. The mud volcanoes are reported not only from the coastal region but also from within the accretionary wedge and the abyssal plain close to it (Wiedicke et al. 2001; Schluter et al. 2002; Delisle 2004; Calves et al. 2010). The mud volcanoes measuring an average diameter of about 1.5 km and up to 65 m high formed as deep as at three kilometer water depth were identified with the help of high resolution echosounding and seismic reflection data (Wiedicke et al. 2001). Wiedicke et al (2001) based on high-resolution sediment-echosounding seaward of the accretionary wedge identified two large zones comprising acoustically transparent facies with diffused scattered reflections marking dome shape features that are 100–300 m wide with very steep and sharp outer boundaries. These transparent mound shaped zones are interpreted as mud volcanoes which are diapirically intruding the overlying wide spread turbiditic facies. A number of smaller transparent zones marking small scale mud volcanoes are associated with the larger structures. The mud volcanoes are interpreted to have developed due to compaction-driven dewatering common in subduction zones with high sedimentation rates. Presence of gas hydrates in the area further facilitates the phenomenon by accelerating the porous, fluid mud movement (Martin et al. 1996; Wiedicke et al. 2001; Delisle 2004).

Study of seismic data from area about 70 km further west of the two recent mud volcanoes as described above, Wiedicke et al. (2001) also reported two ancient buried mud volcanoes about 420 m below sea floor. These structures are 1.5–2 km wide and about 120 m high. The age of these features is estimated as 460 ka BP on the basis of sedimentation rates for the area (Wiedicke et al. 2001). The seismic data shows presence of a number of faults at the base of the both mounds. Calves et al. (2010) further substantiated the formation of ancient mud volcanoes from the Indus submarine Fan by using 2D and 3D seismic reflection data. They recognized a large mud volcanic field comprising nine mud volcanoes, on average 4.5 km wide. Three phases of mud and fluid remobilization are identified in the Early to Middle Miocene, intra-Middle Miocene and in the Late Miocene to Plio-Pleistocene transition. The source of mud was probably of pre-Eocene in age. Similar structures are reported close to the Murray Ridge and from the Indus Fan by Collier and white (1990). However, the rate of sediment supply and tectonic setting of the ridge and fan area are different from the accretionary wedge, suggesting a different triggering mechanism. The formation of mud volcanoes, therefore has been taking place in and around the Makran Subduction Zone since Early Miocene.

6.4 Emergent Mud Islands

The offshore mud volcanoes especially those forming in the shallow shelf area occasionally became emergent islands. Along 700 km long Makran Coast such mud islands have been appearing periodically usually after a major seismic activity in the area. The first published account of such an island was detailed by Sondhi (1947) following emergence of an island off Makran Coast after 8.2 magnitude earthquake on 28 November 1945. The epicenter was along the Makran subduction zone offshore the Makran coast, generating a large tsunami effecting as far as Oman coastal areas to the south-west and Karachi-Bombay coast in east. Four new mud islands appeared at three locations off the Makran Coast following this massive earthquake and were named as Gwadar West Bay Island, Hingol Island and Ormara West Bay Islands (Western Island & Eastern Island) by Sondhi (1947) (Fig. 6.1).

The Gwadar West Bay Mud Island was the largest of the four located about three kilometers from the nearest coast (Fig. 6.1). It was oblong in shape measuring 180×150 m in NNW and SSE direction, about seven meters above the sea-level, composed mainly of pale bluish grey mud which was soft and slippery when wet while stiff and compact on drying. Sondhi (1947) noted rectangular blocks of compact mudstone with extensive boring and encrustation on the north side of the island. These blocks were derived from a single rock unit possibly a compact calcareous mudstone broken *in situ* as it was diapirically emplaced above the water surface. The top surface of the island was covered with compact mud cut by E-W oriented open fissures showing polished surfaces along the shear planes. Fluidized mud and gas periodically bubbled up along the fissures. The Hingol Island (Fig. 6.1) formed only a few kilometers away from the coast measuring about 60 m across and seven meters high above the sea water. It was composed of compact grey mud. Gas was bubbling close to the island gushing sea water up in the air. Near Ormara West Bay two mud islands formed about five kilometer from each other and 16 km offshore from the nearest coast (Fig. 6.1). The Western Island was about one kilometer long and seven meters high whereas the Eastern Island was 1.6–2 km long and over 20 m high. Sondhi (1947) observed a series of N-S oriented parallel ridges and troughs and called these as earth waves. These were slip planes along which the compact mud was sheared as the mud island was diapirically emplaced above the water surface. The mud extruded along one major fracture collapsed to make room for the second wall which eventually resulted into a series of ridges (Delisle 2004). The mud islands were washed out by wave action within months of their emplacement.

More recently new islands have been periodically rising near Makran Coast at more or less the same localities in Hingol and Gwadar area as those of 1945 islands. Twice in recent past mud islands reappeared in the same position of Hingol 1945 island (Fig. 6.1). On 15 March 1999 a mud island named as Malan Island surfaced without any seismic activity in the area. The island was circular in shape and was composed of greenish grey mud breccia and brecciated sandstone. The island was washed by wave action especially during subsequent monsoon season. Delisle et al. (2002) observed minute seafloor hummock and rising gas bubbles through seawater above the island location in November 1999 once the island had completely disappeared. The island reappeared again on 17 November 2010 following a major seismic activity in the area and was named as Sapt Island (Fig. 6.1). The island measured 300×250 m in area with highest point about seven meters above water level (Fig. 6.5a). The island was comprised of light greenish grey mud and large irregular boulders and fragments of sandstone probably broken from the overlying sandy rock units above the fluidized mud (Fig. 6.5b, c). Thick and sticky slurry mud with highly flammable methane gas was spewing from a number of mud vents (Fig. 6.5d). During visit to Sapt Island after four weeks of its appearance, author (D.M. Kakar) observed accumulation of thick viscous slurry closing the vent being burst open by escaping gas pressure (Fig. 6.5c). The viscous mud was thrown many meters up in the air with methane gas that was highly combustible. It was also observed that some vents have been dried out during about past one month as no mud or gas extru-

Fig. 6.5 Mud islands offshore Makran Coast, (**a**) central highest part of Sapt Island appeared in 2010, (**b**) mud shear surfaces as elongated ridges in Sapt Island, (**c**) mud extrusion along new vents after many days of island formation, (**d**) combustible methane gas seepage along the mud vents, (**e**) NASA satellite image of Zalzala Island that appeared near Gwadar Bay in 2013. Mud shear surfaces radiating from the centre of the island, (**f**) general view of Zalzala Island

sion observed in these vents. Apparently the mud becomes thick and gas extrusions diminish resulting in dead vent. The Sapt Island also disappeared after few months due to wave and tide action.

The most recent appearance of a new island was at about the same location of the Gwadar West Bay Mud Island on 24 September 2013 after a Mw = 7.7 earthquake in Awaran area (~380 km northeast of new island), north of Makran coast (Figs. 6.1 and 6.2). This island was named as Zalzala Jazeera (Island) (Fig. 6.5e, f), measuring 18–21 m high, up to 91 m wide, and 37 m long. This island appears very similar to the Sapt Island having viscous mud with angular blocks of compact hard mudstone/sandstone. The mud has good shear strength to be cut in parallel ridges displaced along well defined slip surfaces (Fig. 6.5b). There was a continuous escape of the highly flammable methane gas through a number of vents. This island also disappeared after few months due to wave action.

6.5 Discussion/Mechanism of formation

Fluidized mud flow activity is common along collision zones (Higgins and Saunders 1974), such as Barbados prism (Stride et al. 1982; Westbrook and Smith 1983), the Sunda Arc (Breen et al. 1986; Barber et al. 1986) and Black Sea (Basov and Ivanov 1996). The common driving mechanism is a combination of high sediment accretion rates, presence of under-compacted and over-pressured fine sediments, destabilization of gas hydrates and seismic activity in and around the area (Higgins and Saunders 1974; Henry et al. 1990; Delisle 2004). The Makran Coast is highly suited to formation of mud volcanoes both onshore and offshore where such features are forming since at least Pliocene time. The Hunting Survey Corporation identified a rock unit termed as the "Extrusive Mud Formation" along the Hingol River section comprising similar material and features as most of recent mud volcanoes in the area. Huge amount of gas and water charged mud was extruded to the surface as ancient mud volcanoes emplaced along regional fault structures at least since Pliocene time. Delisle (2004) considers this unit to be part of the Lower Miocene Parkini Formation. This indicates that the mud volcanism was a widespread phenomenon both in past and recent times.

Sediment accretion rates around the Makran prism vary from 110 cm/ka (Wiedicke et al. 2001) to as high as 200 cm/ka (von Rad et al. 1999). This leads to rapid burial of fine grain sediments which were over-pressured and least compacted. Compaction-driven dewatering of sediments is probably the driving mechanism of mud fluidization phenomenon. The mud extrusion is associated with expulsion of flammable methane gas in all mud volcanoes and mud islands along Makran Coast. The bottom simulating reflectors (BSR) studies show existence of widespread gas hydrates in the Makran accretionary prism (Minshull and White 1989). The gas is bacterial (biogenic) in origin (von Rad et al. 2000; Delisle et al. 2002). The gas samples analyzed from various mud volcanoes and Malan Island show fairly similar composition containing over 97 % of microbiologically generated methane (Delisle et al. 2002). The formation of islands during 1945 earthquake was accompanied by flames of self-igniting gas erupted at the mouth of Hingol River "with such a great force that the flames leaped thousands of feet high into the sky" (Sondhi 1947). Such fierce ignition was not reported during reappearance of any of the subsequent islands or mud volcano though gas seeps out of the vents ignites upon exposure to fire. Escaping gas plumes and gas release helps upwelling of fluidized mud at depth.

The distribution of mud volcanoes and mud islands along Makran Coast shows close spatial association with tectonic lineaments such as major folds and faults and

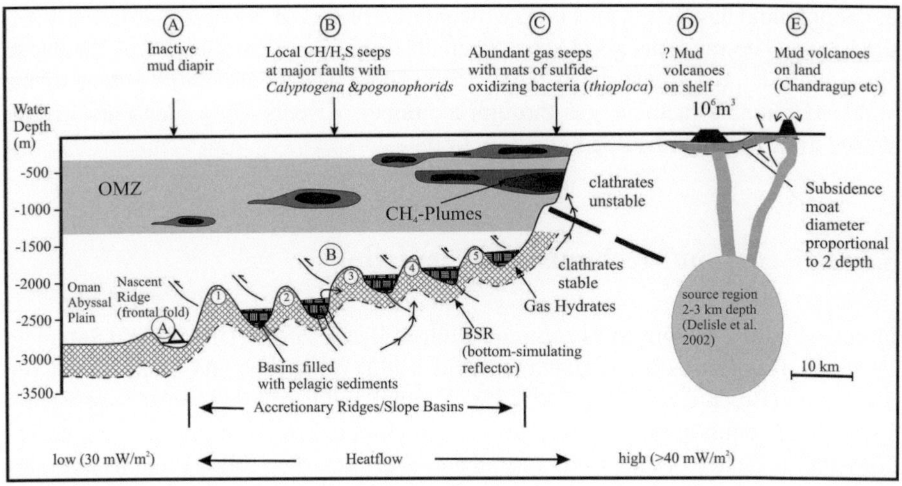

Fig. 6.6 Schematic cross section of the Makran accretionary margin with locations of gas seeps and distribution of methane plumes in water column (after von Rad et al. 2000; Tabrez and Inam 2012; Roger Bilham personal communication). Onshore and offshore mud volcanoes form during methane escape along with fine sediments from a deep source region. OMZ-Oxygen Minimum Zone; BSR-Bottom simulating reflectors; 1–5 are east-west running ridges separated by basins filled with turbidites and pelagic sediments

temporal association with major seismic events in the region. The mud volcanoes are concentrated more along the eastern side of the Makran Coast where deformation is high due to presence of plate scale tectonic lineaments such as Ornach Nal Fault, a splay of the active Chaman Fault defining transform plate boundary between India and Eurasia (Fig. 6.1). Major mud-volcano clusters such as Chandragup, Kandewari and Awaran are all located along major faults and fold axis in the area. This suggests that tectonic lineaments provide an ascent path for mud upwelling. All mud islands formed since 1945 except Malan Island (appeared in 1999) surfaced after a major seismic activity in the region. The 1945 islands of Ormara and Hingol, and 1999 Malan and 2010 Sapt Islands were located close to Ornach Nal fault zone. However, Gwadar Bay Island of 1945 and Zalzala Jazeera of 2013 were not located in close vicinity of any known fault structure with major seismic activity except the Makran Subduction Zone. This indicates mobilization of under-compacted and over-pressured sediments in response to seismic shaking possibly without any role of an adjacent fault. The exceptional case of the Malan Island that emerged when the region was seismically quiet highlights the role of a mechanism other than seismic activity in generating mud islands. We interpret gas hydrates as a source for destabilizing the sediments on seafloor in initiating the fluidized mud movement. Gas hydrates have been identified on the continental slope south of the Makran coast (von Rad et al. 2000), however these become unstable on the continental shelf due to high water temperature and heat flow, and low hydrostatic pressures (Bilham personnel communication) (Fig. 6.6). Pressure release either by decompaction, seismic waves or a local submarine slump may destabilize the solid clathrates shifting to a gas phase rising to the surface with mud and gas bubbles until the local source of clathrates was exhausted. Methane associated with the rising mud plume plays an important role in driving the sediments to the surface by lowering their mean density. Delisle et al. (2002) suggest that the sudden energy release implied by the rapid appearance of the islands is triggered by over-pressured liquids and/or dissolved gases.

6.6 Conclusions

(a) Mud volcanoes and mud islands form both onshore and offshore Makran accretionary wedge. The mud volcanoes cluster along major tectonic lineaments such as anticline axis and active fault zones facilitating ascent path of the fluidized mud.
(b) The mud islands appeared episodically usually after a major seismic activity in the area. The 1945 earthquake of Mw = 8.2 was accompanied by the appearance of four mud islands at three localities near Gwadar, Hingol and Ormara. Successive islands reappeared at Hingol in 1999 as Malan and 2010 as Sapt Islands while near Gwadar a new island reappeared in 2013.
(c) Except for the Malan Island (1999), all other islands surfaced following a major seismic activity in and around the Makran Coast. The Malan Island is therefore interpreted to have been formed by release of methane gas widely distributed in offshore sediments. The compaction of under-compacted and over-pressured fine-grain sediments associated with escaping methane gas therefore play an important role in diapiric movement of the fluidized mud.

References

Barber A, Tjokrosapoetro S, Charlton T (1986) Mud volcanoes, shale diapirs, wrench faults, and mélanges in accretionary complexes, Eastern Indonesia. AAPG Bull 70(11):1729–1741

Basov E, Ivanov M (1996) The Late Quaternary mud volcanism in the Black Sea. Lithol Min Res 31(2):191–197

Bilham R, Lodi S, Hough S, Bukhary S, Murtaza Khan A, Rafeeqi SFA (2007) Seismic hazard in Karachi, Pakistan: uncertain past, uncertain future. Seismol Res Lett 78(6):601–613

Breen NA, Silver EA, Hussong DM (1986) Structural styles of an accretionary wedge south of the Island of Sumba, Indonesia, revealed by Sea MarcII side-scan sonar. Geol Soc Am Bull 97:1250–1261

Byrne DE, Sykes LR, Davis DM (1992) Great thrust earthquakes and aseismic slip along the plate boundary of the Makran subduction zone. J Geophys Res 97(B1):449–478

Calves G, Schwab AM, Huuse M, Van Rensbergen P, Clift PD, Tabrez AR, Inam A (2010) Cenozoic mud volcano activity along the Indus Fan: offshore Pakistan. Basin Res 22(4):398–413

Collier JS, White RS (1990) Mud diapirism within Indus fan sediments: Murray Ridge, Gulf of Oman. Geophys J Int 101:345–353

DeJong KA (1982) Tectonics of the Persian Gulf, Gulf of Oman, and southern Pakistan region. In: Nain AEM, Staehli FG (eds) The ocean basins and margins, vol 6, The Indian Ocean. Plenum Press, New York, pp 315–351

Delisle G (2004) The mud volcanoes of Pakistan. Environ Geol 46:1024–1029

Delisle G, Von Rad U, Andruleit H, Von Daniels CH, Tabrez AR, Inam A (2002) Active mud volcanoes on- and offshore eastern Makran, Pakistan. Int J Earth Sci 91:93–110

DeMets C, Gordon RG, Argus DF, Stein S (1990) Current plate motions. Geophys J Int 101:425–478

Elliot HM (1857) The history of India as told by its own historians: The Muhammadan period In: Dowson J (ed) 3rd edn. London, Trubner. Reprinted AMS Press, New York 1966

Gaedicke C, Schulter H, Roeser HA, Prexl A, Schreckenberger B, Meyer H, Reichert C, Clift P, Amjad S (2002) Origin of the northern Indus Fan and Murray Ridge, Northern Arabian Sea: interpretation from seismic and magnetic imaging. Tectonophysics 355:127–143

Harms JC, Chapel HN, Francis DC (1984) The Makran coast of Pakistan: its stratigraphy and hydrocarbon potential. In: Haq BU, Milliman JD (eds) Marine geology and oceanography of Arabian Sea and Coastal Pakistan. Van Nostr and Reinhold, New York, pp 3–26

Henry P, Le Pichon X, Lallemant S, Foucher J-P, Westbrook G, Hobart M (1990) Mud volcano field seaward of the Barbados accretionary complex: a deep-towed side scan sonar survey. J Geophys Res 95(B):8917–8929

Higgins G, Saunders JB (1974) Mud volcanoes-their nature and origin. In: Jung P, Bolli H, Panchaud R, Saunders J, Schaefer H, Wiedenmayer F (eds) Contributions to the geology and paleobiology of the Caribbean and adjacent areas, vol 84(1). Verhandlungen der Naturforschenden Gesellschaft, Basel, pp 101–152

Hunting Survey Corporation Ltd (1961) Reconnaissance geology of part of West Pakistan. Miracle Press, Ottawa, pp 1–550

Kassi AM, Khan SD, Bayraktar H, Kasi AM (2014) Newly discovered mud volcanoes in the Coastal Belt of Makran, Pakistan-tectonic implications. Arab J Geosci 7(11):4899–4909

Kazmi AH, Abbasi IA (2008) Stratigraphy and historical geology of Pakistan. Department and NCE in Geology, University of Peshawar, Pakistan, 551p

Khan MA, Siddiqui RH, Jan MQ (2010) Temporal evolution of Cretaceous to Pleistocene Magmatism in the Chagai Arc, Balochistan. In: Pakistan. 125th Himalaya-Karakoram-Tibet workshop, San Francisco, USA

Kukowski N, Schillhorn T, Flueh ER, Huhn K (2000) Newly identified strike-slip plate boundary in the northeastern Arabian Sea. Geology 28(4):355–358

Kukowski N, Schillhorn T, Huhn K, von Rad U, Husen S, Flueh ER (2001) Morphotectonics and mechanics of the central Makran accretionary wedge of Pakistan. Mar Geol 173:1–19

Leggett JK, Platt J (1984) Structural features of the Makran fore-arc on Landsat imagery. In: Haq BU, Milliman JD (eds) Marine geology and oceanography of Arabian Sea and Coastal Pakistan. Van Nostr and Reinhold, New York, pp 33–43

Martin JB, Kastner M, Henry P, Le Pichon X, Lallement S (1996) Chemical and isotopic evidence for sources of fluids in a volcano field seaward of the Barbados accretionary wedge. J Geophys Res 101:20325–20345

Minshull TA, White RS (1989) Sediment compaction and fluid migration in the Makran accretionary prism. J Geophys Res 94:7387–7402

Platt JP, Leggett JK, Young J, Raza H, Alam S (1985) Large-scale under plating in the Makran accretionary prism, southwest Pakistan. Geology 13:507–511

Reimann KU (1989) Makran coastal range, Field report no. 8. Hydrocarbon Development Institute of Pakistan, German Advisory Group, Islamabad (BGR Archive No. 0115480)

Schluter HU, Prexl A, Gaedicke C, Roeser H, Reichert C, Meyer H, von Daniels C (2002) The Makran accretionary wedge: sediment thicknesses and ages and the origin of mud volcanoes. Mar Geol 185:219–232

Siddiqui RH (2004) Crustal evolution of the Chagai–Raskoh Arc Terrane, Balochistan, Pakistan. Unpublished PhD thesis, National Centre of Excellence in Geology, University of Peshawar, Pakistan

Snead RE (1964) Active mud volcanoes of Baluchistan, West Pakistan. Geogr Rev 54:546–560

Sondhi VP (1947) The Makran earthquake of 28th November 1945: the birth of new islands. Indian Min 1:154–156

Stride AH, Belderson R, Kenyon N (1982) Structural grain, mud volcanoes and other features on the Barbados ridge complex revealed by GLORIA long-range side-scan sonar. Mar Geol 49:187–196

Tabrez AR, Inam A (2012) Hydrocarbon potential in the Makran offshore area. Search and Discovery Article #80218 (2012)

Tabrez AR, Inam A, Amjad S (1999) Birth of an island "Malan" off Hingol River, Makran, Balochistan. Biogeosciences 8:14–15

von Rad U, Schulz H, Riech V, Den Dulk M, Berner U, Sirocko F (1999) Multiple monsoon- controlled break down of oxygen-minimum conditions during the past 30,000years documented in laminated sediments of Pakistan. Palaeogeogr Palaeoclimatol Palaeoecol 152:129–161

von Rad U, Berner U, Delisle G, Doose-Rolinski H, Fechner N, Linke P, Lückge A, Roeser HA, Schmaljohann R, Wiedicke M, Sonne 122/130 Scientific Parties (2000) Gas and fluid venting at the Makran accretionary wedge off Pakistan. Geo-Mar Lett 20:10–19

Westbrook GK, Smith MJ (1983) Long decollements and mud volcanoes: evidence from the Barbados ridge complex for the role of high pore fluid pressure in the development of an accretionary complex. Geology 11:279–283

White RS, Louden KE (1982) The Makran continental margin: Structure of a thickly sedimented convergent plate boundary. In: Watkins JS, Drake CL (eds) Studies in continental margin geology, vol 34. American Association of Petroleum Geologists Memoir, Tulsa, pp 499–518

Wiedicke M, Neben S, Spiess V (2001) Mud volcanoes at the front of the Makran accretionary complex, Pakistan. Mar Geol 172:57–73

Chapter 7
Response of Sheltered and Built-up Coasts in the Wake of Natural Hazards: The Aftermath of the December 2004 Tsunami, Tamil Nadu, India

Jaya Kumar Seelam and Antonio Mascarenhas

Abstract The geomorphology and the disposition of natural features and type of coast are major factors that control wave heights and govern consequent inundation distances. The connection between tsunami run-up heights and inundation of the hinterland is attempted. Field observations carried out in the aftermath of the December 2004 tsunami reveals that there is no direct relationship between the tsunami run-up height of the incoming wave and the flooding it causes. Coastal sand dunes and forested ecosystems dissipated wave energy during the tsunami event. In comparison, flat and built-up coasts devoid of natural protection suffered maximum damage. Post tsunami restoration initiatives have ignored environmental guidelines.

Keywords Tsunami run-up height • Inundation • Geomorphology • Tamil Nadu

7.1 Introduction

The occurrence of major tsunamis in the recent times has awakened mankind at large to the stark truth that nature has its overpowering strength that humans cannot counter. The Indian Ocean tsunami of December 2004 is one such oceanographic event. The scientific community and the coastal authorities were baffled at the sheer power of violent waves [and the later ones] which devastated large coastal habitations and infrastructure. Our observations indicate that the coastal setting of the southeast Indian coast has by and large played a major role in the inundation and consequent devastation of the coastal regions.

J.K. Seelam (✉) • A. Mascarenhas
CSIR-National Institute of Oceanography, Goa 403 004, India
e-mail: jay@nio.org

V. Santiago-Fandiño et al. (eds.), *Tsunamis and Earthquakes in Coastal Environments*, Coastal Research Library 14, DOI 10.1007/978-3-319-28528-3_7

In this chapter, an attempt is made to look at the coastal setting along part of the southeast Indian landscape with regard to the inundation of the Tamil Nadu coast. Field investigations carried earlier on the December 2004 tsunami run-up heights and inundation limits are revisited. The role and morphology of the coastline are analysed to ascertain the behaviour of sheltered and built-up regions during the tsunami event. The post tsunami scenario and coastal activities over the last decade is also briefly discussed.

7.2 Run-Up Heights and Inundation Limits

Field investigations on the run-up heights and inundation limits along the southeast coast of India along the Tamil Nadu and Pondicherry coasts were carried out by the CSIR-National Institute of Oceanography (Bhattacharya et al. 2005; Ilangovan et al. 2005; Jaya Kumar et al. 2005) amongst others (e.g., Ramanamurthy et al. 2005). The post-tsunami surveys were carried out as per UNESCO-IOC guidelines. From these studies some of the locations with their position, run-up heights, inundation distances and the type of coast are presented in Fig. 7.1 and Table 7.1. Maximum inundation of 862 m was observed at Nagore with the run-up height being about 3.15 m. The inundation at Nagore coast was measured along the region where the coast was un-sheltered with a flat backshore. Though there were casuarina

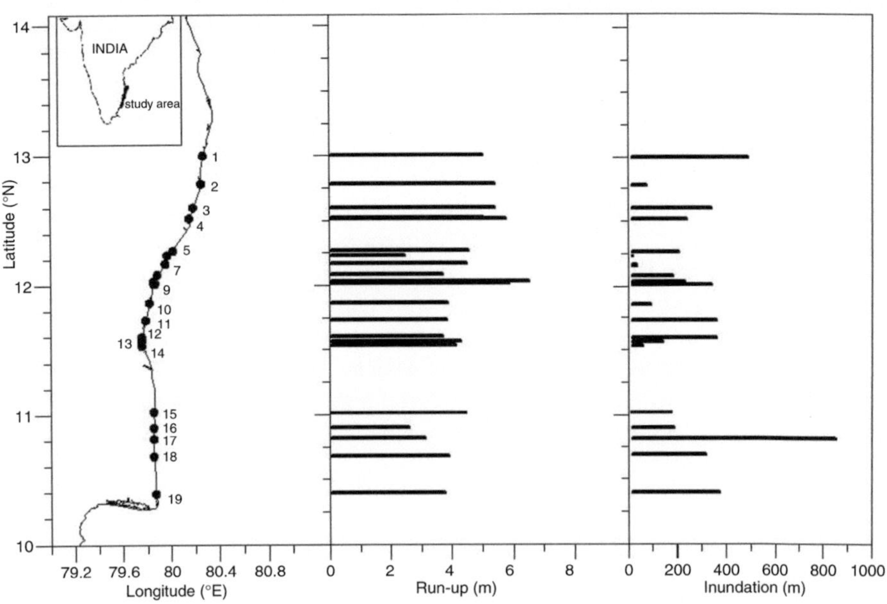

Fig. 7.1 Plot showing surveyed locations, run-up heights and inundation distances (Modified from Jaya Kumar et al. 2005)

Table 7.1 Survey locations, run-up heights, and inundation distances along the coast of Tamil Nadu

	Position	Position	Run up	Inundation	Type of coast	
Station Name	Latitude °N	Longitude °E	(m)	(m)	Coast region	Backshore
1. Besant Nagar Chennai	13°00'35.28"	80°16'34.05"	5	113	Dune coast	Wooded
2. Kovalam	12°47'19.09"	80°15'16.02"	5.4	90	Rocky; dune	Steep
3. Mahabalipuram	12°36'55.05"	80°11'56.08"	5.4	349	Un-sheltered	Flat backshore
4. Sadarangapattinam	12°31'22.09"	80°09'56.28"	5.8	250	Un-sheltered	Flat backshore
5. Alambarai Kuppam	12°16'04.92"	80°00'55.02"	4.6	216	Dune coast	High backshore
6. Kottai Kadu Kuppam	12°14'42.00"	79°58'59.04"	2.5	31	Dune coast	Steep backshore
7. Ekkiar Kuppam	12°10'49.00"	79°57'37.03"	4.5	49	Dune coast	Steep backshore
8. Nochi Kuppam	12°05'08.46"	79°53'55.01"	3.7	196	Dune coast	Flat backshore
9. Periyakalapet (North)	12°02'03.18"	79°52'17.07"	6.5	240	Dune coast	High back dunes
Periyakalapet (Middle)	12°01'40.08"	79°52'02.52"	5.4	175	Built-up	Steep backshore
Periyakalapet (South)	12°01'27.18"	79°51'56.08"	5.9	350	Un-sheltered	Hind dunes
10. Pudu Kuppam	11°52'03.00"	79°49'09.36"	3.1	105	Un-sheltered	Steep backshore
11. Thevanampattinam	11°44'40.00"	79°47'18.09"	3.9	370	Built-up	Flat backshore
12. Nanjalingampettai	11°36'45.03"	79°45'34.08"	3.7	372	Dune coast	High dune
13. Reddiar Pettai	11°34'34.00"	79°45'28.06"	3.2	155	Dune coast	Steep backshore
14. Velangirayan	11°32'15.04"	79°45'44.07"	3.7	324	Dune coast	Low flat backshore
15. Tarangambadi	11°01'37.02"	79°51'21.00"	4.4	160	Built-up	Flat backshore
16. Karaikal	10°54'50.03"	79°51'08.04"	2.6	200	Un-sheltered	Flat backshore
17. Nagore	10°48'47.52"	79°51'04.38"	3.1	862	Un-sheltered	Flat backshore
18. Velangani	10°40'48.68"	79° 51'02.22"	3.9	325	Un-sheltered	Flat backshore
19. Arukattuthurai	10°23'30.51"	79°52'07.14"	3.8	385	Un-sheltered	Flat backshore

Note that a low wave height at Nagore induced a large inundation as compared to Periyakalapet for example

plantations on both the southern and northern side of the beach access, the inundation took place along the middle stretch of the beach which was un-sheltered. The coastal approach road to the beach served as tsunami pathway and resulted in the one of the largest measured inundation distances for this particular tsunami.

A maximum run-up height was measured at the northern portion of the Periyakalapet beach along the Pondicherry coast with a run-up height of 6.54 m. The beach at Periyakalapet had mixed coastal setup with a low dune on the southern end, buildings and habitation in the middle and high back dune and coconut and palm grove towards the northern region. The run-up heights observed within a kilometre distance between the north and south regions of the Periyakalapet showed a difference of about 1.2 m and the difference in inundation limits varied by 175 m.

Some intriguing differences in the inundation distances for similar run-up heights were noted from the measurements. For example, for a run-up height of about 4.6 m the inundation distance was about 216 m at Alambaraikuppam; whereas for similar run-up height of 4.5 m at Ekkiarkuppam had an inundation of only 49 m. At Tarangambadi, the inundation distance was about 160 m even though the run-up height was 4.4 m. This observation of similar run-up heights resulting in different inundation distances is attributed to both the beach face being un-sheltered or not as well as to the backshore region being dune region or having steeper slope.

The run-up heights are shown to be directly proportional to the incoming wave height for a given beach slope; this surmise includes tsunami waves as well (e.g., Gedik et al. 2005; Synolakis 1987). Therefore, the inundation distances are also proportional to the run-up height (or wave height) for a given un-obstructed beach slope. Consequently, higher the run-up height for a given location, the tsunami wave height should in fact have been proportionally higher and a greater inundation distance. However, in the case of Tamil Nadu coast for example, our field surveys show that wherever the run-up height of the incoming wave was high the inundation distance was, at times, nominal. The opposite was also found to be true. This observation implies that there is no direct or positive correlation between tsunami wave height and the inundation distance.

A similar phenomenon was noted as far as inundations due to storm surges are concerned. For example, a surge of 6.0 m observed during October 1963 at Cuddalore, December 1964 at Rameshwaram, and October 1971 at Paradip resulted in inundations up to 10–25 km only during the 1971 cyclone that hit Paradip coast (see Mascarenhas 2004). It is obvious that coastal features did control the flooding of the hinterland.

Therefore, it can be confidently confirmed that the relation between tsunami wave height and run-up distances are influenced additionally by some other factors. Obviously, "some" natural feature such as the local topography or the geomorphological disposition of the coast plays a definite role by acting as a controlling factor in the wake of oceanographic events.

It is pertinent to note that several other natural and man-made features that comprise a coast also influenced the advance of the tsunami onshore. For example, (a) breaches on/along dune faces influenced wave advance inland; such examples are plenty and often mark areas where local inhabitants access the beach, (b) roads and parking lots that abut the beach facilitated the propagation of the seawater further; this was observed at Nagapattinam where the wave overtopped the beach and moved about 800 m unhindered, (c) creeks and rivers played a major role in allowing high waves to travel unchallenged; this phenomenon was witnesses at Karaikal where the columns of a heritage bridge collapse due to impact of powerful waves. Similarly, coastal roads served as pathways for the tsunami advance, one such example is at Nagore where the inundation was a maximum.

7.3 Sheltered Coasts and Built Up Coasts: Functional Aspects

The concept of sheltered and built-up coasts as attempted in this paper needs to be defined. Any sandy beach is fully exposed to the vagaries of the ocean. It is what lies behind the beach and the morphology of the backshore, from the high tide line towards the hinterland, that determines whether a coast presents sheltered characteristics or not. Therefore, a well formed high dune with associated vegetation, a flat back shore devoid of vegetal cover, or a sea front lined by a series of dwellings would all react differently in the wake of an oncoming wave.

The beaches along the north Tamil Nadu coast e.g., along Chennai, Mahabalipuram and Sadarangapattinam (Kalpakam) coasts have a steeper beach face and varying backshore steepness. Sand dunes exist in the hinterland all along the coast which is covered by habitation at most of the locations. South of Chennai up to about 40 km, the coastline is occupied with the built-up area less than 100 m from waterline. Similar coastal setting is observed at almost all locations in the study region. Exceptions being smaller hamlets located behind large dune fields e.g., Nanjalingampettai.

Extensive surveys supported by measurements and observational analyses of that damage caused by the tsunami reveal that natural coasts indeed possess inherent protective value. For example, geomorphic features such as coastal sand dunes and associated vegetation assumed importance as far as the attenuation of the incoming wave is concerned. Similarly, the extensive casuarinas forests that mark the coast of Tamil Nadu also played a defensive role in absorbing the violent waves. To confirm this observation, two sites along Tamil Nadu coast, Nanjalingampettai in the central part and Velangani further south have been selected. Beach profiles from the water line to the uppermost accessible point are presented in Figs. 7.2 and 7.3.

The coast at Nanjalingampettai (Fig. 7.4) is characterized by prominent sand dunes more than 5 m high and very steep seaward gradients as seen in the beach profile. Luxuriant casuarina forests interspersed with dense coconut groves characterize coastal vegetation along this region. As determined by our surveys, the wave up-rush was as high as 3.7 m. It is evident that the wave stopped at the dune, the dune crest being higher. Casuarina plantations that mark this area further aided a complete attenuation of the wave onslaught. The dense vegetation provided the requisite obstruction to the tsunami flow thereby dissipating its energy and stopping the tsunami flow further inland.

Several similar situations can be cited. For example, Besant Nagar in Chennai on the northern region of Tamil Nadu coast, Silladi and Samanthanpettai along the southern region of Tamil Nadu coast are some examples where villages behind casuarina forests remained secure. Therefore, it is pertinent to note here that dwellings behind such geomorphic features and located behind the dense plantations benefited from natural protection. Natural forested landforms such as well-preserved dunes and casuarina forests acted as natural biological shields by saving life and property. Neither there was damage to any habitation nor any loss of life for loca-

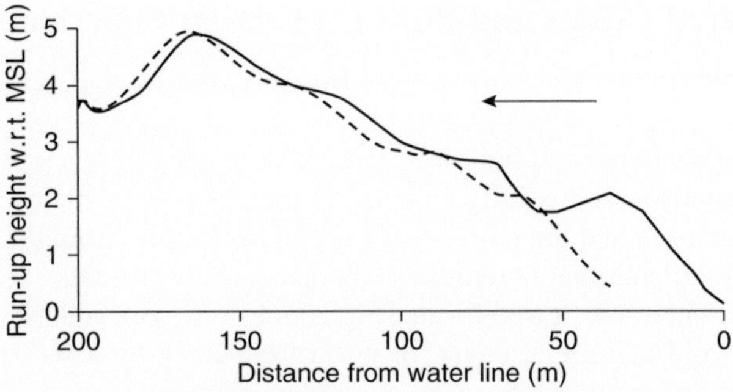

Fig. 7.2 Profile at Nanjalingampettai beach during April 2005 (*solid line*) and January 2006 (*dotted line*). The *arrow mark* shows the measured tsunami run-up height (Source: Mascarenhas and Jaya Kumar 2008)

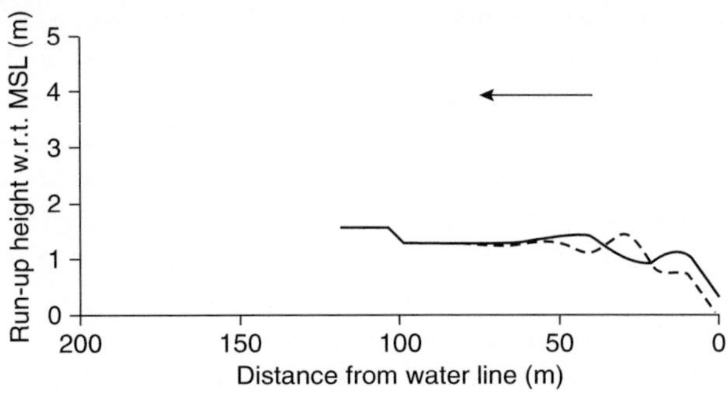

Fig. 7.3 Profile at Velangani beach during April 2005 (*solid line*) and January 2006 (*dotted line*). The *arrow mark* shows the measured tsunami run-up height (Source: Mascarenhas and Jaya Kumar 2008)

tions behind the forested dune complexes as houses behind densely vegetated high dunes are all intact.

The beach at Velangani is around 80 m wide but presents a flat character. There is hardly any presence of frontal dunes as these have been mutilated due to trampling. No vegetation exists on the lee side of the beach. Beach profiles at different times over a period of 1 year showed that very little beach building takes place. Only about 0.4 m berm built up was seen over a period of 1 year. The berm location at Velangani varies with season and about 20 m variation was observed within a year as observed from the measurements.

Before the tsunami, the entire coastal strip, right up to the beach, was occupied by dense make-shift structures occupied by a large fishing community (Fig. 7.5). This strip was probably a dune field, but no evidence is found at present. Coastal geomorphology was obliterated, as sand dunes were flattened in favour of houses. Here, the wave run-up of 3.9 m bypassed the flat beach thus razing whatever came its way. As they lacked natural protection, all the structures were smashed.

Fig. 7.4 Beach region at Nanjalingampettai showing a prominent sand dune with dense vegetation. The village behind this dune was not damaged by tsunami

Fig. 7.5 Photo of a flat beach at Velangani that is continuously overused (Note that the huts and temporary huts have existed at this pilgrimage site for a long time (since 1998 at least), as seen in the accompanying photograph. These structures were wiped off in totality. Loss of life was very high)

Similarly, complete destruction along the coast of Tarangambadi and a kilometre long Akkaraipettai – Keechankuppam sector in particular was noted. Having been dangerously close to the sea, the direct hit eliminated these structures in totality. Maximum loss of life was encountered at Akkaraipettai where more than 6.000 inhabitants perished. Around 900 people lost their lives at Velangani.

In brief, the two sites presented above are in fact coastal villages with contrasting coastal geomorphology, the first being a village with an intervening forest and the second, a cluster of dense structures along a flat open sea front. Therefore, coastal vegetated landforms in the form of natural elevation and natural vegetation was found to be of prime importance in significantly neutralizing incoming wave energy. The issue of protective elevation in combination with protective vegetation is the key issue here.

Our observations are in conformity with other research conducted elsewhere. It is shown that areas with tree cover were markedly less damaged than areas without (Danielsen et al. 2005; Olwig et al. 2007). The villages on the coast were completely destroyed, whereas those behind the mangrove suffered no destruction even though the waves damaged areas unshielded by vegetation north and south of these villages. Mangroves and casuarina plantations attenuated tsunami-induced waves and protected shorelines against damage from tsunami by reducing wave amplitude and energy (Danielsen et al. 2005).

The role of mangroves in attenuating wave energy was proved earlier as research had also contended that Sunderbans mangrove of Bengal suffer less from wind and surges (Clark 1996). The defensive role of mangroves during cyclones as natural barriers protecting the life and property of coastal communities has been demonstrated at Bhitarkanika on the northern Indian east coast (Badola and Hussein 2005), and by post-cyclone imageries of Orissa coast (Nayak et al. 2001).

7.4 Post Tsunami Scenario: A Decade Later

Having realized that coastal plantations were indeed beneficial as bio-shields, massive plantation programs were initiated, both at the government level as well as by local communities. One year later, significant rebuilding activity was noted along the entire coast of Tamil Nadu, as noted during our surveys in January 2006. The entire stretch from Chennai to Mahabalipuram was covered by aggressive plantation programs. Thousands of casuarina trees were planted on the bare dunes impacted by the tsunami. Unfortunately, at several sites as in Silladi, casuarina saplings were planted too on the beach, sometimes by leveling dunes which the tsunami had only partially damaged. The well intentioned plantation programs were however conducted haphazardly. Hinterland vegetal species were planted along the beach. Being too close to the coastline, it is doubtful if these saplings will survive. Cases of inappropriate methods of afforestation are plenty. Obviously, the selection of species for coastal afforestation lacks prudent planning. Table 7.2 describes the overall scenario since the December 2004 tsunami event.

Table 7.2 Characteristics of some the locations surveyed along Tamil Nadu coast, from north to south, where post-tsunami damage studies were carried out. Beach width (April 2005) is measured from the water line mostly to the base of dune (wherever identified). High dunes refer to prominent geomorphic features. Damage to trees refers to the strip of damaged vegetation, measured physically, mostly from the dune line. Figure 'zero' signifies that damage was nil

Coastal site	Characteristics/state of beach, dune and vegetation prior to December 2004 tsunami	Run-up (m)	Damage of coastal vegetation strip in April 2005 (m)	Damage to dwellings and infrastructure	State of vegetation/ forests (January 2006)	Present scenario (December 2013)
Besant Nagar, Chennai	Inland trees on high dunes	5.0	0	No damage to dwellings	Trees survived in totality; dune creepers have returned	The beach dune system is by and large intact
Southern part	Dunes razed; huts on beach		–	Seafront huts washed off	No evidence of vegetation	The huts of fisherfolk are back
Vada Nemelli	High dunes capped by vegetation; extensive casuarina forests	–	13–25	This coastal strip was uninhabited	Only frontal trees perished; casuarinas have revitalized; uprooted dune plants are now growing normally	Several new bungalows have sprouted right along the dune field
Mahabalipuram	A tourist spot, flattened dunes, scattered trees	5.4	0	The temple wall was partly damaged	Patches of casuarinas are still seen	A boulder wall now occupies the northern part of coast
Periyakalapet	Shorefront dwellings within coconut groves	5.3–6.5	0		All coconut trees survived wave attack	
Thevanampattinam	Small dwellings within coconut groves	3.8	0	Almost all houses suffered heavy damage	No damage to coconut trees	The row of houses appears to have been abandoned; a rubble mound sea wall is built here

(continued)

Table 7.2 (continued)

Coastal site	Characteristics/state of beach, dune and vegetation prior to December 2004 tsunami	Run-up (m)	Damage of coastal vegetation strip in April 2005 (m)	Damage to dwellings and infrastructure	State of vegetation/ forests (January 2006)	Present scenario (December 2013)
Nanjalingampettai	High dunes with coconut plantations; casuarinas are wide spread	3.7	10–17	Houses behind dune fields remained intact	Sea facing strip of casuarinas suffered nominal damage; bent trees are sprouting	The village continues with normal life
Poompuhar	A pilgrimage site, marked by tall palm trees and coconut groves	2.2	0	Some damage to existing retaining walls	Existing 19 palm trees were found unharmed	New sea wall is built
Vanagiri	Historical shore temples; sparse vegetation	–	–	Some monuments were dislodged and collapses	Few existing trees survived	This coast suffered erosion; a sea wall is now built here
Tarangambadi	Dense dwellings along the shore	1.9–3.4	0	Frontal houses were reduced to rubble	Whereas all dwellings were flattened, Few existing trees remain	A rubble mound sea wall is built here
Karaikal, north	A tourist spot, sparse vegetation	2.6	0	Nominal damage to the park	No damage to trees	The park is rebuilt
South of creek	Luxuriant casuarina forests	–	~15	No loss of property	Casuarina forest is intact except for bent trees within the frontal strip	Casuarina forest exists as before
Nagore	Houses, huts of fisher folk amidst dense coconut trees	–	0	Several houses collapsed	Except few, coconut trees however remained intact	Rebuilding activity was noted
Silladi	High vegetated dunes; extensive casuarinas plantations	1.7–3.1	7–19	No damage to infrastructure	Only frontal trees were bent and stripped of bark; rest are surviving	The flat back shore is now occupied by a lagoon

Samanthanpettai	Conspicuous dunes with vegetation	1.9–3.3	5	Only a few frontal houses collapsed	Casuarinas were partly affected; dune vegetation has recuperated	Some houses are rebuilt
Nambiar Nagar	Prominent, high sand dunes	–	0	Hardly any damage was noted	Backshore trees behind dunes are intact	Some houses are rebuilt
Nagapattinam lighthouse	Flattened dunes, huts,	–	0	Frontal wall was damaged at foundations	Nominal damage to a few trees	The same wall has been fortified
Akkaraipettai – Keechankuppam	A thickly populated village on the coast; scattered trees	–	–	Severe damage to dwellings; houses washed off entirely; maximum human loss was reported from this village	Severe damage to dwellings; a patch of casuarinas still standing; scattered coconut trees survived	The former village appears like a 'ghost town'; a rubble mound sea wall is built here
Kallar	A low, flat, flood prone area	–	0	The area took the form of a lagoon	Few shrubs were uprooted	–
Velangani	A pilgrimage site; flat beach with thatched huts and make-shift restaurants	2.1–4.7	–	Structures washed off in totality; backshore plants are still standing	Structures washed off in totality; backshore plants are still standing	All the makeshift structures are back; it is business as ususal at this site
Puatadi	A pristine coast with well formed, high vegetated sand dunes; thriving casuarinas forests	–	0	The coast is free of any houses	Casuarina trees remained unaffected; presently growing normally	–

The extreme impact was the complete destruction of Akkaraipettai, where heavy loss of life and property was recorded. Panic stricken villagers started planting several rows of coconut trees all over the kilometer long beach. This activity can be termed as the most unscientific plantation program ever undertaken. One year later in January 2006, an erosive scarp that formed destroyed all the saplings. This is a clear case of a lack of appropriate administrative, management and plantation guidelines.

It is confirmed that the profusely vegetated stretches of Tamil Nadu coast displayed an exceptional resilience by dissipating high waves. Our field measurements confirmed that only the frontal strip of casuarina woodland ranging from 0 to 25 m were twisted, bent and stripped of leaf cover. These observations reiterate that coastal vegetation in general and casuarina trees in particular played an essential role as efficient biological buffers against powerful oceanographic episodes (Mascarenhas 2004, 2006; Danielsen et al. 2005; Mascarenhas and Jaya Kumar 2007, 2008).

However, a question that is being raised is the levels of natural protection that a dense woody coastal plantation can offer and is now a matter of debate. Experimental studies however show that mangrove forests attenuate wave energy. The rate of dissipation of waves depends on the density of forests and the diameter of tree roots and trunks (Massel et al. 1999). Currents thus flow around rather than through the forest (Furukawa et al. 1997).

In comparison, empirical evidence for casuarina as a bio-shield is definitely lacking. Due to this issue, an evaluation of sand dune flora from the Indian coasts was undertaken (Rodrigues et al. 2011). A three layered biozone of creepers – shrubs – taller trees, from the pioneer (fore) dune – mid dune – hind dune was proposed. Extensive casuarina plantations established in the 1990s and earlier as a cyclone protection measure along the Orissa coast was ineffective in preventing damage; these plants are known to bear a shallow root system and easily uproot with strong wind and beach erosion (Schmid et al. 1993). The use of C. equisetifolia is seen to be inappropriate in the foredune and mid-shore areas as they are known to pose a threat to marine fauna (Schmid et al. 1993) and may obstruct the natural succession patterns of vegetation (Rodrigues et al. 2011). Presently, hardly any effort has been made to validate the role and functions of bio-shields from a scientific perspective.

Our field surveys along the Tamil Nadu coast during January 2006 and December 2013 indicated active (re)construction activity along the coast. Rapid reconstruction of some houses, on the same foundations, was noticed at some places. Similar situations were observed near Chennai, Periyakalapet, Thevanampattinam, Tarangambadi and parts of Nagapattinam. The worst scenario was noted at Velangani: all the makeshift restaurants, thatched shops and shacks, washed off in totality in December 2004 were back in January 2006, and standing even today. It is business as usual at this pilgrimage site. Similarly, the tsunami had over topped the sea wall at Nagapattinam port. The one at Karaikal was also uprooted over a length of around 250 m. Despite adverse impacts, the rubble mound seawall at Pondicherry city is widened, and the one at Nagapattinam port is being rebuilt in concrete, at a high cost, as a taller and longer fortification.

The coast of Tamil Nadu has witnessed plenty of building activity in the form of new ports and harbors. This infrastructure has necessitated the fixing of coast perpendicular groynes at several places, apparently to keep waterways free of siltation. The observed impact of this activity is the rapid erosion of beaches on the downstream side. One such example is the withdrawal of the beach at Pondicherry. Global research has proved the multiple impacts of sea walls on the coast; sandy beaches in front of hard structures get narrower, sand often vanishes, erosion is exacerbated and the area becomes deeper (Fletcher et al. 1997; Carmo et al. 2010).

Despite scientific research that is available, and without giving a thought to the consequences of erosion control measures, the coastal authorities went about building sea walls as a tsunami protection option. Massive rubble mound shore parallel structures now occupy long stretches of Tamil Nadu coast. By placing the rubble mound shore parallel seawall for coastal protection, the coastal erosion problem is given a boost instead of alleviating it. Hard protection measures have been shown to increase erosion along the coast causing problems on the down drift side of the sediment transport region. Hard protection measures in the inter-tidal region have been shown to increase erosion along the coast causing problems on the down drift side of the sediment transport region (e.g. Pilkey and Wright 1988; Krauss and McDougal 1996). The effect of seawalls on the beaches is not immediately felt in this region, as it is seen that the effects of seawalls on the coast is a long term feature. Figure 7.5 shows the positioning of a 2.5 m high rubble mound sea wall at Thevanampattinam. As a consequence, coast seems to have lost its sand as no beach is visible even at low tide as can be confirmed from the accompanying photograph (Figs. 7.6 and 7.7).

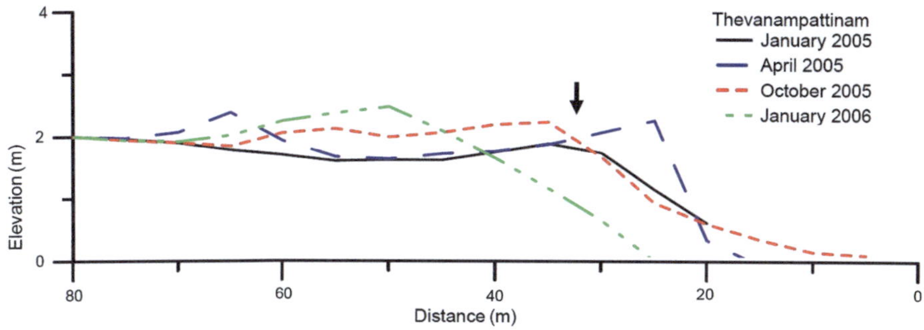

Fig. 7.6 Beach profiles at Thevanampattinam, one of the sites where post tsunami measurements were carried out. The arrow indicates the location of shore protection structure placed on the beach. The site can no longer be used for beach profile studies (Profiles source: Jaya Kumar et al. 2008)

Fig. 7.7 Photo of the beach at Thevanampattinam (December 2013) is now occupied by a rubble mound sea wall. The approximate location of the sea wall is marked on the profile (Note that no beach is visible along the sea wall even at low tide)

7.5 Conclusions

The belief that the wave run-up is proportional to the flooding of hinterland during extreme oceanographic events was found to be untrue in the case of December 2004 that affected the coast of Tamil Nadu. Coastal geomorphology is the key factor that controls both, the run-up heights and the inundation that may (or may not) follow. Prominent geomorphic features such as dune fields remarkably stalled the power of the violent tsunami waves. Similarly, dense plantations mostly in the form of casuarina forests performed extraordinarily by blocking the onrushing sea water. Thus, elevated coastal sand dunes and dense plantations, either independently or jointly, dissipated wave energy. Villages behind such natural coasts remained intact. In comparison, houses along the sea front suffered maximum damage as they lacked natural protection. Post tsunami rebuilding, restoration and management initiatives show that (a) new dwellings are erected at the same place along vulnerable sites, often without obtaining mandatory approvals and by disregarding coastal regulations; (b) plantation programs which are in fact to be appreciated, but have not followed proper guideline, as evident from haphazard cultivation styles; (c) it appears that long strips of coast are now fixed with concrete and rubble mound sea walls which are constructed without complying with the role of coastal processes; regular monitoring will reveal the efficacy of these structures and the sustainability of the beaches which at the moment are already under the grip of erosive processes; and, as such (d) most of the activities have overlooked long term environmental impacts.

Acknowledgements Authors acknowledge the support and infrastructure provided by the CSIR-National Institute of Oceanography during the field surveys carried out by the authors. The post-tsunami survey team members of CSIR-NIO are duly acknowledged for their help in obtaining the beach profiles. Prof. Hitoshi Tanaka, Tohoku University, Japan and Prof. V. Sundar, IIT-Madras, Chennai, India, are acknowledged for their initiative and support in making this manuscript. This manuscript (Ref. No. 8085) is cleared by the Publication Committee of CSIR-NIO.

References

Badola R, Hussain SA (2005) Valuing ecosystem functions: an empirical study on the storm protection function of Bhitarkanika mangrove ecosystem, India. Environ Conserv 32:85–92

Bhattacharya GC, Ilangovan D, Jaya Kumar S, Manimurali R, Kocherla M, Alagarsamy R, Gowthaman R, Naik KA (2005) Observations of post tsunami reconnaissance investigations along eastern coastal tract of India following the devastating tsunami of 26th December 2004, NIO; Dona Paula, Goa 403 004; India., NIO/TR-08/2005; 2005; 109 pp

Carmo JA, Reis CS, Freitas H (2010) Working with nature by protecting Sand Dunes: lessons learned. J Coast Res 26:1068–1078

Clark JR (1996) Coastal zone management handbook. Lewis Publication, Boca Raton

Danielsen F, Sørensen MK, Olwig MF, Selvam V, Parish F, Burgess ND, Hiraishi T, Karunagaran VM, Rasmussen MS, Hansen LB, Quarto A, Suryadiputra N (2005) The Asian tsunami: a protective role for coastal vegetation. Science 310:643

Fletcher CH, Mullane RA, Richmond BM (1997) Beach loss along armored shorelines on Oahu, Hawaiian Islands. J Coastal Res 13:209–215

Furukawa K, Wolanski E, Muller H (1997) Currents and sediment transport in mangrove forests. Estuar Coast Shelf Sci 44:301–310

Gedik N, Irtem E, Kabdasli S (2005) Laboratory investigation on tsunami run-up. Ocean Eng 32:513–528

Ilangovan D, Jaya Kumar S, Gowthaman R, Tirodkar G, Ganeshan P, Naik GN, Manimurali R, Michael GS, Ramana MV, Naik KA (2005) Inundation, run-up heights, cross-section profiles and littoral environment along the Tamil Nadu coast after 26th December 2004 Tsunami, NIO; Dona Paula, Goa 403 004; India., NIO/TR-03/2005; 2005; 71 pp

Jaya Kumar S, Ilangovan D, Naik KA, Gowthaman R, Tirodkar G, Naik GN, Ganeshan P, Manimurali R, Michael GS, Ramana MV, Bhattacharya GC (2005) Run-up and inundation limits along the southeast coast of India due to the great Indian Ocean Tsunami. Curr Sci 88(11):1741–1743

Jaya Kumar S, Naik KA, Ramanamurthy MV, Ilangovan D, Gowthaman R, Jena BK (2008) Post-tsunami changes in littoral environment along southeast coast of India. J Environ Manag 89:35–44

Krauss NC, McDougal WG (1996) The effects of seawalls on the beach: Part I, an updated literature review. J Coast Res 12(3):691–701

Mascarenhas A (2004) Oceanographic validity of buffer zones for the east coast of India: a hydrometeorological perspective. Curr Sci 86:399–406

Mascarenhas A (2006) Extreme events, intrinsic landforms and humankind: post-tsunami scenario along Nagore—Velankanni coast, TamilNadu. Curr Sci 90:1195–1201

Mascarenhas A, Jayakumar S (2007) Protective role of coastal ecosystems in the context of the tsunami in Tamil Nadu coast, India: implications for hazard preparedness. In: Murty TS, Aswathanarayana U, Nirupama N (eds) Indian ocean tsunami. Taylor & Francis, London, pp 423–435

Mascarenhas A, Jayakumar S (2008) An environmental perspective of the post-tsunami scenario along the coast of Tamil Nadu, India: role of sand dunes and forests. J Environ Manag 89:24–34

Massel SR, Furukawa K, Brinkman RM (1999) Surface wave propagation in mangrove forests. Fluid Dyn Res 24:219

Nayak SR, Sarangi RK, Rajawat AS (2001) Application of IRS-P4 OCM data to study the impact of cyclone on coastal environment of Orissa. Curr Sci 80:1208–1213

Olwig MF, Sørensen MK, Rasmussen MS, Danielsen F, Selvam V, Hansen LB, Nyborg L, Vestergaard Parish F, Karunagaran VM (2007) Using remote sensing to assess the protective role of coastal woody vegetation against tsunami waves. Int J Remote Sens 28:3153–3169

Pilkey OH, Wright HL (1988) Seawalls versus beaches. J Coast Res SI-4:41–64

Ramanamurthy MV, Sundaramoorthy S, Pari Y, Ranga Rao V, Mishra P, Bhat M, Usha T, Venkatesan R, Subramanian BR (2005) Inundation of sea water in Andaman and Nicobar Islands and parts of Tamil Nadu coast during 2004 Sumatra tsunami. Curr Sci 88(11):1736–1740

Rodrigues RS, Mascarenhas A, Jagtap TG (2011) An evaluation of flora from coastal sand dunes of India: rationale for conservation and management. Ocean Coast Manag 54:181–188

Schmid JL, Addison DS, Donnelly MA, Shirley MA, Wibbels T (1993) The effect of Australian pine (Casuarina equisetifolia) removal on loggerheads turtle (Caretta caretta) incubation temperatures on Keewaydin island Florida. J Coast Res 55:214–220

Synolakis CE (1987) The runup of solitary waves. J Fluid Mech 185:523–545

Chapter 8
Characteristics of Shoreline Retreat Due to the 2011 Tohoku Earthquake and Tsunami and Its Recovery After Three Years

Keiko Udo, Kaoru Tojo, Yuriko Takeda, Hitoshi Tanaka, and Akira Mano

Abstract We investigate the characteristics of shoreline retreat due to the 2011 Tohoku Earthquake and Tsunami and its recovery after 3 years over a wide area from Aomori Prefecture to Ibaraki Prefecture by analyzing the data of crustal deformation, tsunami height, and aerial photographs. There were no significant retreats where the seawall remained unbreached and where river mouths were distantly located regardless of the tsunami overflow depth over the seawall, indicating that the seawall condition during tsunami strikes is one of the most important factors that prevent serious shoreline retreat. After 3 years from the event, recovery of the eroded Ria coasts has been insignificant compared to the plain coasts. The key natural factors for the recovery are the upward crustal deformation after the earthquake and sediment supply from adjacent coasts or rivers after the tsunami; however, these could not be expected at the Ria coasts at least in near future.

Keywords Coastal land loss • Sediment transport • Coastal structure • Satellite image • Beach erosion

8.1 Introduction

At 14:46 JST on 11 March 2011, a magnitude 9.0 earthquake occurred off the Pacific Coast of Miyagi Prefecture, causing devastating tsunami waves (Fig. 8.1a). Crustal deformation due to the earthquake induced land subsidence, and the maximum crustal deformation measured by permanent GPS stations was an east-southeast deformation of 5.3 m and a downward deformation of 1.2 m at the Ojika

K. Udo (✉) • Y. Takeda • A. Mano
International Research Institute of Disaster Science, Tohoku University, Sendai, Japan
e-mail: udo@irides.tohoku.ac.jp

K. Tojo • H. Tanaka
Department of Civil and Environmental Engineering, Tohoku University, Sendai, Japan

V. Santiago-Fandiño et al. (eds.), *Tsunamis and Earthquakes in Coastal Environments*, Coastal Research Library 14, DOI 10.1007/978-3-319-28528-3_8

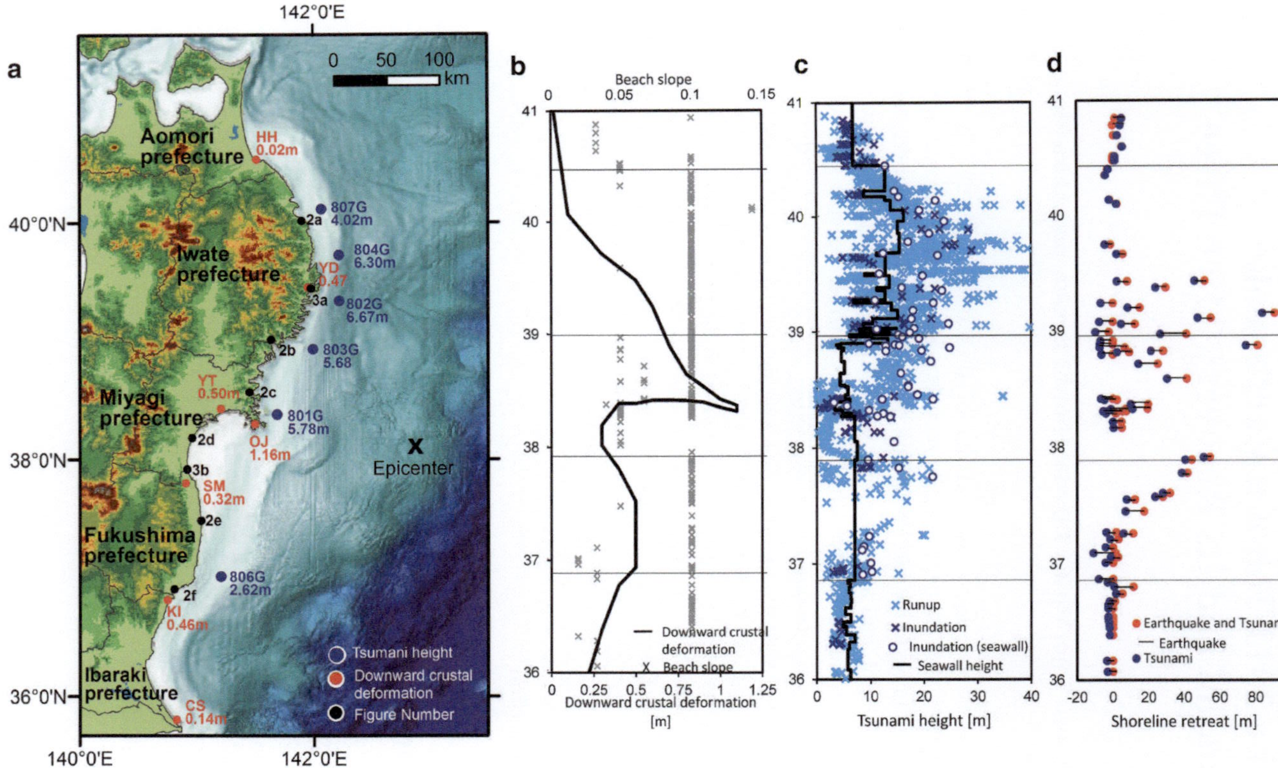

Fig. 8.1 Distributions of (**a**) epicenter of the earthquake, permanent GPS stations where downward crustal deformation was measured (Geospatial Information Authority of Japan 2013), and GPS buoys where tsunami height observed (Kawai et al. 2011), (**b**) downward crustal deformation observed by permanent GPS stations and beach slope (Biodiversity Center of Japan 2012), (**c**) tsunami watermark height (TTJS 2012) and seawall height before the earthquake and tsunami, and (**d**) shoreline retreat due to the earthquake (*red closed circle*) and tsunami (*blue closed circle*). *Black circles* from 2a to 2f indicate the locations given in Figs. 8.2 and 8.3 respectively (e.g. 2a refers to Fig. 8.2a Fudaigawa; 3a refers to Fig. 8.3a Funakoshi)

Table 8.1 Vertical crustal deformation before and after the 2011 earthquake

Place	1–11 March, 2011 (due to the earthquake)	11 March to 17 April, 2011 (1 month after the earthquake)	11 March 2011 to 11 April 2013 (2 years after the earthquake)
Hachinohe (HH)	−0.02 m	+0.01 m	+0.01 m (less than 0.001 m month^{-1})
Yamada (YD)	−0.47 m	−0.08 m	+0.02 m (less than 0.001 m month^{-1})
Yamoto (YT)	−0.51 m	+0.05 m	+0.18 m (0.007 m month^{-1})
Soma (SM)	−0.32 m	+0.02 m	+0.11 m (0.004 m month^{-1})
Kitaibaraki (KI)	−0.46 m	+0.01 m	+0.10 m (0.004 m month^{-1})
Choshi (CS)	−0.15 m	+0.05 m	+0.06 m (0.002 m month^{-1})

Source: Geospatial Information Authority of Japan (2013)

peninsula (OJ in Fig. 8.1a; Geospatial Information Authority of Japan 2013; Kinoshita and Takagishi 2011). The downward deformation estimated to be more than 0.5 m extended widely in the coastal areas from Yamada to Ishinommaki (Yamoto) and from Namie to Nakoso (Table 8.1 and Fig. 8.1b).

The tsunami was observed with GPS buoys installed off the coast of Iwate, Miyagi, and Fukushima prefectures (Fig. 8.1a; Kawai et al. 2011); afterward, it inundated broadly from Hokkaido Prefecture to Chiba Prefecture as pointed out by the 2011 Tohoku Earthquake Tsunami Joint Survey (TTJS) Group 2012 (Hereinafter referred to as TTJS 2012). The maximum tsunami crest reached the coasts of Ofunato with a tsunami height of more than 8.0 m at 15:18, Ishinomaki more than 8.6 m at 15:25, Miyako more than 8.5 m at 15:26, and Soma more than 9.3 m at 15:51 (Kawai et al. 2011). The run-up height greater than 10 m distributed along 530 km of coast in direct distances and the height greater than 20 m along 200 km of coast (Fig. 8.1b; TTJS 2012; Mori et al. 2012). The total tsunami inundation area from Aomori Prefecture to Chiba Prefecture was 561 km^2 (Geospatial Information Authority of Japan 2013). In the Sendai Plain, it reached up to 5 km inland from the coast line.

One of the most serious coastal damages due to the earthquake and tsunami was coastal land loss. It was reported that significant beach shoreline retreats of several hundred meters and consequent coastal land losses, especially in the vicinity of river mouths and seawall breaches, extended over a wide area from Iwate Prefecture to Fukushima Prefecture (Public Works Research Center 2011; Nagasawa and Tanaka 2012), causing serious problems in the reconstruction process. There have been a number of studies on the characteristics of local erosion due to the earthquake and tsunami (e.g., at the Misawa coast in Aomori Prefecture by Nakamura et al. 2012; the Sanriku coasts in Iwate Prefecture by Kato et al. 2012 and Okayasu et al. 2013; the Sendai coasts in Miyagi Prefecture by Tanaka et al. 2012; Udo et al. 2012; Adityawan et al. 2014, and Udo et al. 2015; the Nakoso coast in Fukushima Prefecture by Sato et al. 2014; and the Kashima coast in Ibaraki Prefecture by

Kuriyama et al. 2014 and Sato et al. 2015). However, few studies have been made on the overall characteristics of the shoreline change and recovery over a wide area by focusing on coastal areas regardless of the severity of erosion.

This study aims to clarify the characteristics of shoreline retreat due to the earthquake and tsunami and its recovery after 3 years in a wide area from Aomori Prefecture to Ibaraki Prefecture by analyzing crustal deformation data, tsunami height data, and aerial photographs before and after the earthquake and tsunami.

8.2 Method

The study area and the points where crustal deformation data were acquired are shown in Fig. 8.1a. The crustal deformation data before and after the earthquake, the tsunami height data, and aerial photographs before and after the earthquake and tsunami were analyzed also identifying the relationships among them. The crustal deformation data were obtained at coastal areas of Hachinohe, Yamada, Yamoto, Soma, Kitaibaraki, and Choshi in the period of 1 March 2011 to 11 April 2013 by Geospatial Information Authority of Japan (2013) (Table 8.1 and Fig. 8.1b).

The tsunami dataset utilized in the study refers to the tsunami inundation height (watermark height) data observed by TTJS (2012) (Fig. 8.1c). As the dataset includes some data with low reliability, only those highly reliable ("*clear watermark*" or "*no clear watermark but reliable information from witnesses*") were used. The dataset of seawall height before the earthquake and tsunami occurred was obtained from official records from Aomori (2014), Iwate (2011), Miyagi (2011), Fukushima (2011), and Ibaraki (2012) prefectures. These datasets had been used to calculate the tsunami overflow depth over the seawall and relate it in turn to the seawall breach (Ministry of Land, Infrastructure, Transport and Tourism, 2011) and consequent coastal erosion. The overflow depth was calculated by subtracting the seawall height from the inundation height.

In order to investigate the shoreline retreat from Aomori Prefecture to Ibaraki Prefecture where the tsunami inundation was observed, available aerial photographs from 1947 to 2013 provided by Geospatial Information Authority of Japan, satellite images since 2000s on Google Earth, and coastline change dataset provided by Biodiversity Center of Japan (2012) were used. The coastline change data before the tsunami occurred were extracted from satellite images (IKONOS or GeoEye-1) obtained from November 2000 to January 2010, while those after the tsunami took place were extracted from aerial photographs obtained by Geospatial Information Authority of Japan in the period from 12 March 2011 to 12 April 2012. The data were collected at an interval of 50 m in coastal length and were corrected according to the tidal levels. From the existing dataset, we extracted only the shoreline change data in beach areas where the tsunami height was obtained by TTJS. The consistency of the shoreline change data was in turn checked with the aerial photographs and satellite images whereby only reliable information was extracted. The shoreline changes were averaged for each beach having a width between 0.1 and 10.0 km and thereafter classified into two sets i.e. changes caused by the downward crustal deformation due to the earthquake and by sediment transport due to the tsunami. The changes caused by the crustal deformation were calculated by dividing the downward deformation by the coastal slope. Likewise, the changes caused by the sediment transport were calculated by subtracting the one due to the crustal deformation from the total shoreline change. The coastal slope data was provided by Biodiversity Center of Japan (2012) (Fig. 8.1b).

8.3 Results

8.3.1 Shoreline Retreat Due to the 2011 Earthquake and Tsunami

Distributions of the downward crustal deformation, tsunami and seawall heights as well as shoreline retreat are shown in Fig. 8.1b–d, respectively. Examples of typical erosions in the vicinity of river mouths are shown in Fig. 8.2; and those in the vicinity of seawall breaches are shown in Fig. 8.3.

Significant retreats were observed mainly in the vicinity of river mouths and seawall breaches from Iwate Prefecture to Fukushima Prefecture, but there were no significant retreats where the seawall remained unbreached and where river mouths were distantly located regardless of the tsunami overflow depth. Meanwhile, there was no clear relationship between the magnitude of the downward crustal deformation and that of the shoreline change because the beach slopes at Ria coasts were roughly 0.1 which is relatively steep compared to those at plain coasts of 0.05, although some presented large downward deformation of more than 0.5 m. The coastal erosion due to the crustal movement was insignificant compared to that due to the tsunami; in other words, the coastal erosion was mainly caused by the tsunami sediment transport.

At the Fudaigawa and Kesengawa (Rikuzen-takata) rivers shown in Fig. 8.2a, b located in Iwate Prefecture (also as shown in black circles in Figs. 8.1a and 8.2a, b), the shoreline retreated by up to 100 m at the Fudaihama beach by the river's mouth (Fig. 8.2a) (Seo and Ohkushi 2012); while up to 500 m at the Takata beach by the Kesengawa river mouth (Fig. 8.2b) (Kato et al. 2012).

The changes at the Kitakamigawa and Natorigawa rivers in Miyagi Prefecture (Fig. 8.2c, d) went up to 1200 m at the Suka beach at the Kitakamigawa's river mouth (Fig. 8.2c), while the 200 m wide left sand barrier located at the Natorigawa river mouth was almost fully eroded (Fig. 8.2d) (Udo et al. 2012) although the damage done to the right barrier was relatively small.

The Koitogawa and Samegawa rivers are located in Fukushima Prefecture (Fig. 8.2e, f). The 200 m wide sand bar at the former river mouth was found to be totally eroded (Fig. 8.2e), while the 300 m wide sand bar at the latter mouth had the same fate (Fig. 8.2f) (Adityawan et al. 2012).

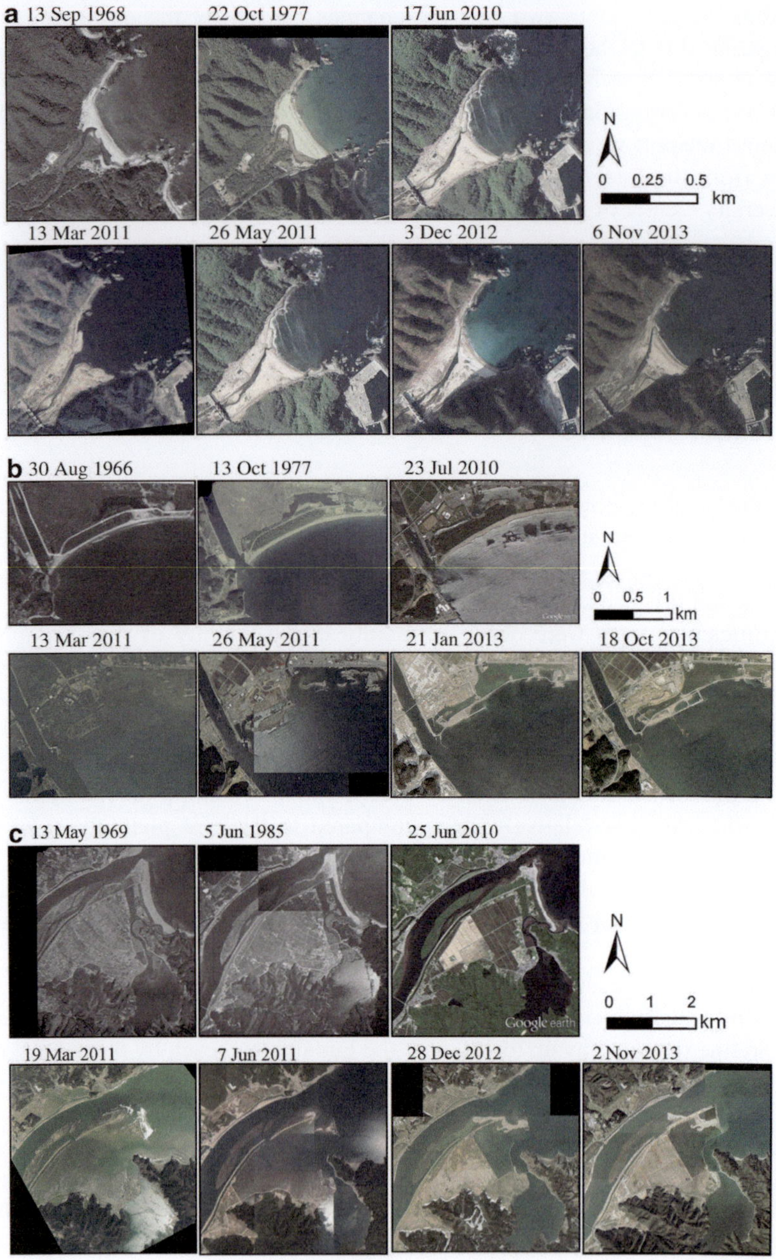

Fig. 8.2 Aerial photographs in the vicinity of river mouths of the (**a**) Fudaigawa, (**b**) Kesengawa, (**c**) Kitakamigawa, (**d**) Natorigawa, (**e**) Koitogawa, and (**f**) Samegawa rivers from 1960s to 2013 (provided by the Geospatial Information Authority of Japan); See Fig. 8.1a *(No aerial photographs exist for the black images)*

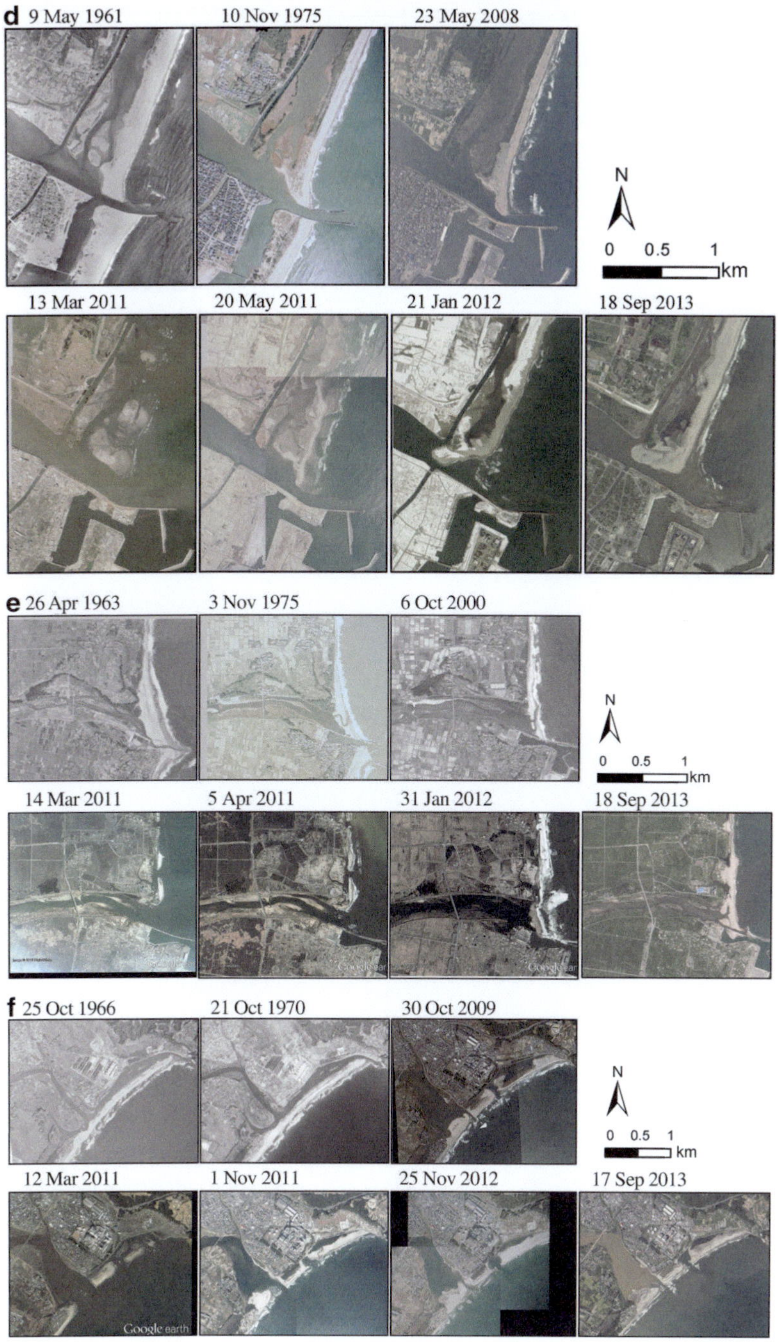

Fig. 8.2 (continued)

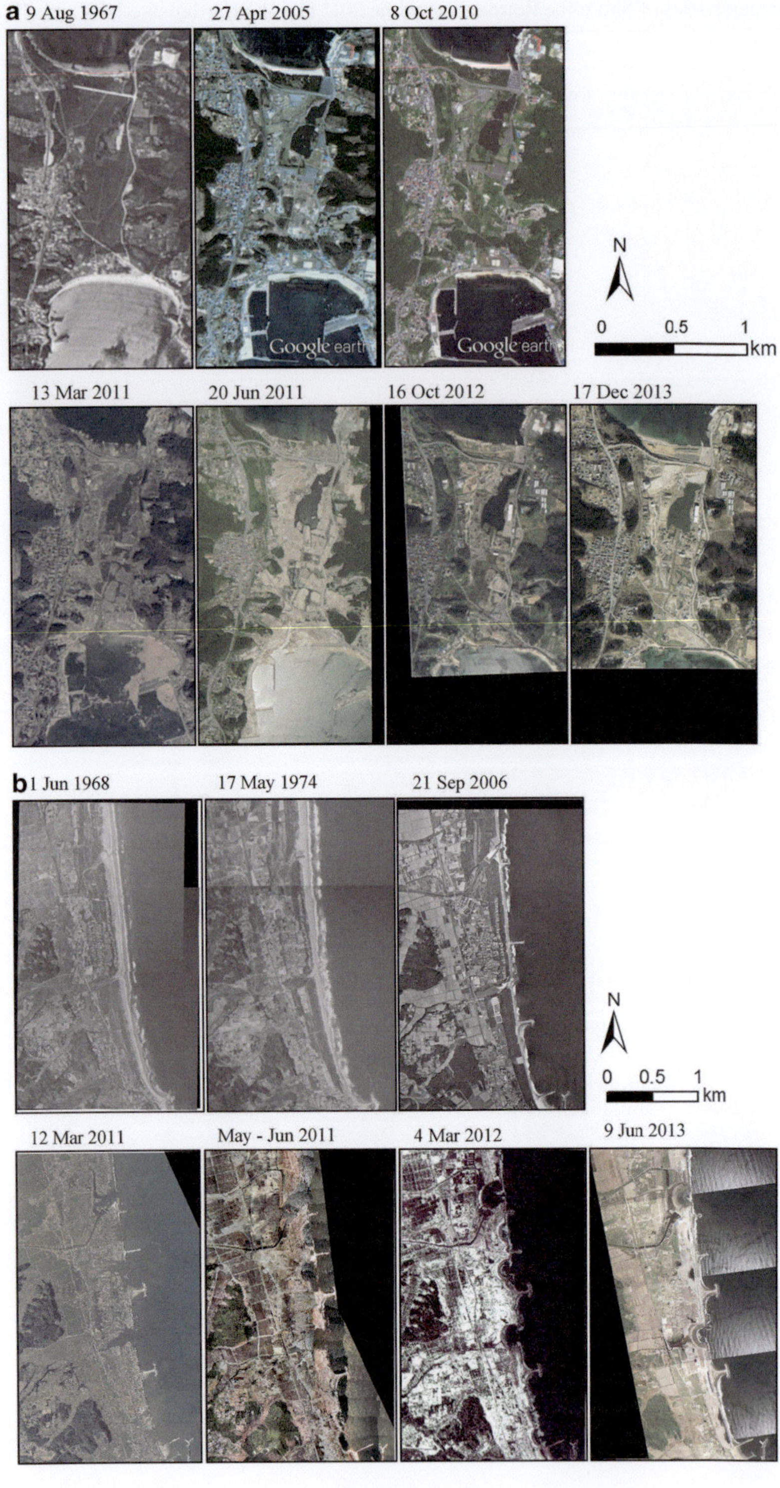

Fig. 8.3 Aerial photographs of (**a**) Funakoshi and (**b**) Yamamoto beaches from 1960s to 2013 (Geospatial Information Authority of Japan 2013; see Fig. 8.1a (*No aerial photographs exist for the black images)*

The significant retreats at the Takata (Fig. 8.2b) and Suka (Fig. 8.2c) beaches were likely caused by both the downward deformation and sediment transport; however, in the other cases the retreats were mainly caused by sediment transport.

It is important to note that the Uranohama and Tanohama beaches, which are tombolo beaches of the Funakoshi peninsula (Fig. 8.3a), were likewise significantly eroded because the tsunami crossed them in north-south direction. Although the Yamamoto beach shown in Fig. 8.3b is not located in the vicinity of river mouth, it also suffered a significant retreat by up to 300 m due to sediment transport around the seawall breaches (Udo et al. 2015).

8.3.2 *Three-Year Shoreline Change After the Earthquake and Tsunami*

After 3 years from the event, several shorelines along the coast have been recovering:

(a) Fudaihama beach (Fig. 8.2a); the shoreline recovered by approximately 30 m after a year; and thereafter it has almost recovered to its original position.
(b) Natorigawa river mouth (Fig. 8.2d); a new barrier was formed again after a half year but 150 m inland on the left side from its original position while the shoreline advanced 50 m after 3 years.
(c) Koitogawa river mouth (Fig. 8.2e); a new bar was formed again but 100 m inland of its original position after half year while the shoreline advanced 50 m after 2 years becoming stable afterwards.
(d) Samegawa river mouth (Fig. 8.2f); a new sand bar was formed again after a year but 100 m inland from its original position while the shoreline advanced 50 m after 3 years.
(e) Yamamoto beach (Fig. 8.3b); the most eroded inland section recovered rapidly, up to 100 m after a half year and further increasing by 50 m after a year but finally reaching 200 m after 3 years, however this also indicates that there is a slowdown in the speed of the recovering process.

On the other hand, the shoreline recovery was insignificant at the Takata (Fig. 8.2b), Suka (Fig. 8.2c), Uranohama, and Tanohama beaches (Fig. 8.3a). The reasons could be the fact that these beaches are located at Ria coastlines, where the downward deformation was larger and sediment supply has been estimated to be limited as the beach length is relatively short and river mouths are distantly located.

8.4 Discussion and Conclusions

Based on the existing studies (e.g., Nagasawa and Tanaka 2012; Udo et al. 2015) and the results of this study, the mechanism of the shoreline retreat due to the earthquake and tsunami can be summarized as follows: the shoreline changes due to the earthquake at first depending on the magnitudes of the vertical crustal deformation and the coastal slope, and then due to the tsunami depending on the seawall damage. Both the downward crustal deformation and coastal slope are related to the shoreline retreat due to inundation, while the overflow depth is related to the seawall damage consequently causing local erosion in the vicinity of the seawall breaches. In addition, sand bars at the river mouths are eroded due to the backwash from the river to the sea.

The key natural factors affecting the recovery in the above-mentioned sites are the vertical crustal deformation after the earthquake and sediment supply from adjacent coasts or rivers after the tsunami. If the upward deformation occurs or sediment is supplied from adjacent coasts or rivers, the eroded coast is expected to recover depending on the magnitudes of upward deformation, beach slope, and sediment supply, and vice versa. The recovery of the eroded beach where the upward deformation does not occur and the sediment is not supplied is expected to be difficult.

Acknowledgments This research was supported by the Special Project Research of IRIDeS (International Research Institute of Disaster Science, Tohoku University) and JSPS KAKENHI Grant Number 26289158.

References

Adityawan MB, Tanaka T, Nagabayashi H (2012) Investigation of shoreline change caused by the Great East Japan Tsunami 2011 around Samegawa River mouth. Proc Int Sess Coast Eng Conf 3:31–35

Adityawan MB, Dao NX, Tanaka H, Mano A, Udo K (2014) Morphological changes along the Ishinomaki Coast induced by the 2011 Great East Japan Tsunami and the relationship with coastal structures. Coast Eng J 56(3):1450016

Aomori Prefecture (2014) http://www.pref.aomori.lg.jp/soshiki/kendo/kasensabo/files/2013-0129-1301.pdf. Accessed June 2015

Biodiversity Center of Japan (2012) 2012 Survey of the natural environment in the Pacific coast of Tohoku district. http://www.shiokaze.biodic.go.jp/PDF/h24report/h24_report_all.pdf. Accessed June 2015 (in Japanese)

Fukushima Prefecture (2011) http://www.pref.fukushima.lg.jp/uploaded/attachment/10267.pdf (in Japanese)

Geospatial Information Authority of Japan (2013) Crustal movements of permanent GPS stations observed by GEONET System. http://www.gsi.go.jp/chibankansi/chikakukansi40005.html. Accessed June 2015 (in Japanese)

Ibaraki Prefecture (2012) http://www.pref.ibaraki.jp/bukyoku/doboku/01class/class06/kaigan/tsunamisinnsui/file/02mezasubeki/mezasubeki.pdf. Accessed June 2015 (in Japanese)

Iwate Prefecture (2011) http://www.pref.iwate.jp/dbps_data/_material_/_files/000/000/008/326/teibou2.pdf. Accessed June 2015 (in Japanese)

Kato F, Noguchi K, Suwa Y, Sakagami T, Sato Y (2012) Field survey on tsunami-induced topographical change. J Japan Soc Civil Eng Ser. B3 (Ocean Eng) 68(2):I_174–I_179 (in Japanese)

Kawai H, Satoh M, Kawaguchi K, Seki K (2011) Characteristics of the 2011 off the Pacific coast of Tohoku Earthquake Tsunami. Report Port Airport Res Inst 50(4):1–64 (in Japanese)

Kinoshita S, Takagishi M (2011) Generation and propagation of static displacement estimated using KiK-net recordings. Earth Planets Space 63:779–783

Kuriyama Y, Takahashi K, Yanagishima S, Tomita T (2014) Beach profile change at Hasaki, Japan, caused by 5-m-high tsunami due to the 2011 off the Pacific coast of Tohoku Earthquake. Mar Geol 355:234–243

Ministry of Land, Infrastructure, Transport and Tourism (2011) Reconstruction measures for seawall and revetment. http://www.mlit.go.jp/river/shinngikai_blog/kaigantsunamitaisaku/dai02kai/dai02kai_siryou5.pdf. Accessed Jun 2015 (in Japanese)

Miyagi Prefecture (2011) Report on the first year after the 2011 Tohoku Earthquake and Tsunami. http://www.pref.miyagi.jp/uploaded/attachment/40635.pdf. Accessed June 2015 (in Japanese)

Mori N, Takahashi T, The 2011 Tohoku Earthquake Tsunami Joint Survey Group (2012) Nationwide post event survey and analysis of the 2011 Tohoku Earthquake Tsunami. Coast Eng J 54:1250001

Nagasawa T, Tanaka H (2012) Study of structural damages with massive geomorphic change due to tsunami. In: Proceedings of the 8th international conference on disaster management, pp 165–170. www.iiirr.ucalgary.ca/files/iiirr/A6-3_.pdf

Nakamura Y, Nishimura Y, Putra PS (2012) Local variation of inundation, sedimentary characteristics, and mineral assemblages of the 2011 Tohoku-oki tsunami on the Misawa coast, Aomori, Japan. Sediment Geol 282:216–227

Okayasu A, Shimozono T, Yamazaki H, Nagai T, Sato S (2013) Severe erosion of sandbar at Unosumai river mouth, Iwate, due to 2011 Tohoku Tsunami. Proc Coastal Dyn 2013:1311–1320

Public Works Research Center (2011) Tsunami report. http://pwrc-nagisa.jp. Accessed June 2015 (in Japanese)

Sato S (2015) Seawall performance along Southern Coast of East Japan impacted by the 2011 Tohoku Tsunami; a note for the reconstruction process. In: Santiago-Fandiño V, Kontar Y, Kaneda Y (eds) Post-tsunami hazard reconstruction and restoration. Springer, Cham, pp 191–209

Sato S, Okayasu A, Yeh H, Fritz HM, Tajima Y, Shimozono T (2014) Delayed survey of the 2011 Tohoku Tsunami in the former exclusion zone in Minami-Soma, Fukushima Prefecture. Pure Appl Geophys 171:3229–3240

Seo N, Ohkushi K (2012) Preliminary research of tsunami deposits around the mouth of the Fudai river, Iwate Prefecture. Bull Grad School Hum Dev Environ Kobe Univ 5(2):157–164

Tanaka H, Tinh NX, Umeda M, Hirao R, Pradjoko E, Mano A, Udo K (2012) Coastal and estuarine morphology changes induced by the 2011 Great East Japan Earthquake Tsunami. Coast Eng J 54:1250010

The 2011 Tohoku Earthquake Tsunami Joint Survey (TTJS) Group (2012) Results of tsunami surveys. http://www.coastal.jp/tsunami2011/. Accessed June 2015

Udo K, Sugawara D, Tanaka H, Imai K, Mano A (2012) Impact of the 2011 Tohoku Earthquake and Tsunami on beach morphology along the Northern Sendai Coast. Coast Eng J 54:1250009

Udo K, Takeda Y, Takamura M, Mano A (2015) Serious erosion of the southern Sendai coast due to the 2011 Tohoku earthquake tsunami and its recovery process. In: Santiago-Fandiño V, Kontar Y, Kaneda Y (eds) Post-tsunami hazard reconstruction and restoration. Springer, Cham, pp 225–236

Chapter 9
Investigating the 2011 Tsunami Impact on the Teizan Canal and the Old River Mouth in Sendai Coast. Miyagi Prefecture; Japan

Mohammad Bagus Adityawan and Hitoshi Tanaka

Abstract This study investigates the effect of old river mouth and the Teizan Canal to the 2011 tsunami at Arahama and Akaiko, both part of Sendai Coast, Japan. At both locations, the Teizan Canal is situated parallel to the shoreline. This canal affected the tsunami inundation as well as the return flow, effectively draining the water back to the sea. In addition, there were old river mouths, which have been closed for many years in these locations. Nevertheless, the return flow of the 2011 tsunami caused the breaching of the sandy coast at the old river mouth locations. A detail topography analysis was carried out based on DEM data. The topography data show that this area is a river basin with mild slope towards the old river mouth location. The data was further analyzed by performing a flow accumulation analysis using GIS. The results confirm that the topographical features in both locations caused the flow to accumulate through the local channel and canal towards the locations of the old river mouth, which caused the breaching.

Keywords Japan Earthquake and Tsunami 2011 • River mouth • Breaching • Teizan Canal

9.1 Introduction

The north east coast of Japan was greatly affected by The Great East Japan Earthquake and Tsunami on 11 March 2011. The massive tsunami wave, generated by the 2011 Earthquake, reached the affected shoreline and caused destruction to the coastal

M.B. Adityawan (✉)
Water Resources Engineering Research Group, Institut Teknologi Bandung, Jalan Ganesha 10, Bandung 40132, Indonesia

Department of Civil Engineering, Tohoku University 6-6-06 Aoba, Sendai 980-8576, Japan
e-mail: mbagus_st@yahoo.com; bagus.adityawan@ftsl.itb.ac.id

H. Tanaka
Department of Civil Engineering, Tohoku University 6-6-06 Aoba, Sendai 980-8576, Japan
e-mail: tanaka@tsunami2.civil.tohoku.ac.jp

V. Santiago-Fandiño et al. (eds.), *Tsunamis and Earthquakes in Coastal Environments*, Coastal Research Library 14, DOI 10.1007/978-3-319-28528-3_9

area. Estuaries were also severely damaged in many places (Tanaka et al. 2014a). The tsunami wave was reported at a maximum runup height of approximately 40 m in rocky coasts (Mori and Takahashi 2011). The Sendai Plain was also severely hit by the tsunami (Adityawan et al. 2012). The wave caused a large number of casualties as well as damages to coastal structures along the coast (Supasri et al. 2012).

The tsunami caused tremendous damages to the coastal structures although some were purposely built against tsunamis. Tappin et al. (2012) concluded that the protection around the Sendai coastline may have not been highly effective in case massive tsunamis stroke as in the event of 2011. Therefore, the recovery process requires a re-evaluation and re-design of these structures in the face of future tsunami events. The impacts on sea fences by tsunamis can be evaluated and quantified using conventional engineering approaches leading to new designs (Shito et al. 2014; Adityawan et al. 2014).

After the 2011 tsunami event it was found that elevated roads, embankments, vegetations, as well as the local topographical features acted as countermeasures against the tsunami albeit with different efficacy. Another important lesson from this event was observed in the Sendai Plain where the Teizan Canal, parallel to shoreline, which also contributed to reduce the tsunami's impact as already published by Dao et al. (2013). The authors concluded that a canal parallel to shoreline could be an effective countermeasure against tsunami waves, however it depends on the tsunami's magnitude. Likewise, Tsimopulou et al. (2012) concluded that canals alongside other types of coastal protection could provide a multiple defense line against tsunamis.

As result of the studies, the concept of using canals as protection features have been widely studied and proposed in many places. Usman et al. (2014) proposed a multiple defense line in Indonesia based on local topography and vegetation using a numerical model. He concluded that the hollow topography and an embankment could act as an empty channel when running in parallel to the shoreline hence effectively protecting the coastal area from tsunami waves.

A multiple defense line may provide an effective defense against an incoming tsunami; however, if breached it is important to have an effective drain against the tsunami-induced inundation. Based on our observations the Teizan Canal, along with a local channel, showed to collect the return flow of the 2011 tsunami by also causing the breaching of the sandy coastline by the old river mouth. Tanaka et al. (2012) shown the effectiveness of this phenomenon as an efficient drainage system for the tsunami's return flow since the inundation time was also reduced. Although this author has also made studies on the tsunami propagation in rivers (Tanaka et al. 2014b) still the mechanism on how the canal as well as the old river mouth act to the tsunami return flow is unclear; nonetheless, it is important to consider this particular approach as a part of the disaster mitigation plan.

Unfortunately, numerical as well as physical models often omit channels and canals in their components due to their small size in the affected areas; hence the need for further detailed research. This study aims to investigate the effect of the canals and old river mouths to the tsunami return flow in the 2011 tsunami focusing on Arahama and Akaiko locations in Sendai; the return flow mechanism was analyzed using historical map and GIS analysis of topographic data.

9.2 Study Area

The study area focusing on Arahama and Akaiko coastlines is shown in Fig. 9.1. Both locations were greatly affected by the 2011 tsunami although most of their shorelines remained relatively unharmed even though breaching occurred in some places. Figure 9.2 shows a video snapshot, captured from a helicopter during the tsunami at Akaiko where it becomes clear that a very strong return flow occurred through the canal also causing breaching on the sandy coastline. Satellite images, captured immediately after the tsunami shows the local breaching due to the tsunami at Arahama and Akaiko (Figs. 9.3 and 9.4 respectively). It is estimated that the breaching width at Arahama was approximately 50 m while the one at Akaiko 130 m approximately.

9.2.1 Analysis of Historical Maps

The Teizan Canal (constructed in 1597) along with the Touna and Kitakami Canals are a series of canals stretching for approximately 46 km from the Abukuma River to the Kitakami River (Fig. 9.5). All the canals were constructed over the span of almost 290 years for different purposes (Encyclopedia of the Teizan Canal) and are still maintain and kept operational until today (Fig. 9.5 inbox photo). The more recently constructed Kitakami and Touna Canals (1878–1881 and 1883–1884, respectively) are about 20–60 m wide with a maximum water depth of approximately 0.8 m; both were mainly constructed to support the development of a port at Nobiru while the Teizan Canal was constructed to support the development of Sendai City.

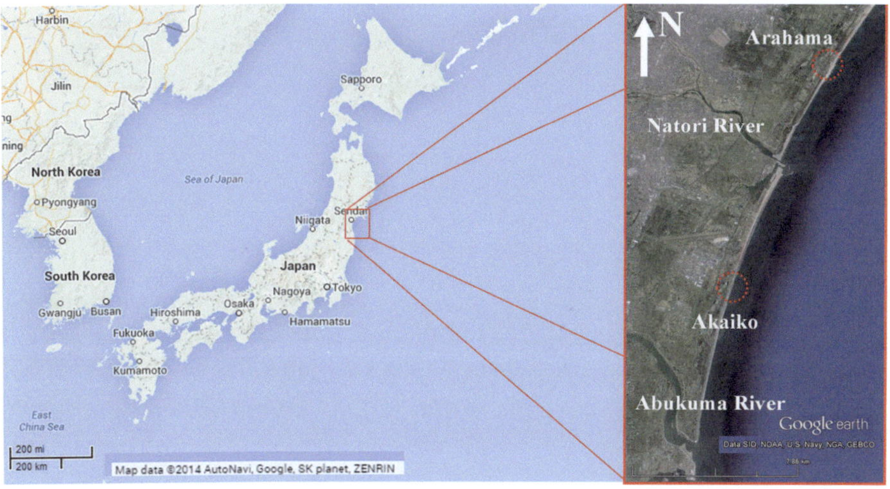

Fig. 9.1 Study area; Sendai. Japan

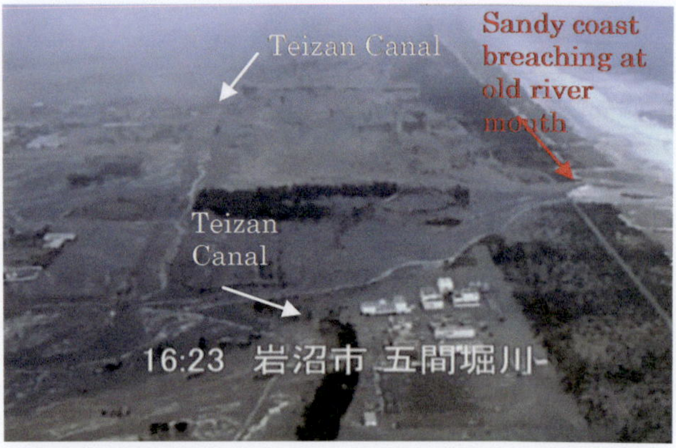

Fig. 9.2 Video snapshot at Akaiko during the 2011 tsunami (Tohoku Regional Development Bureau)

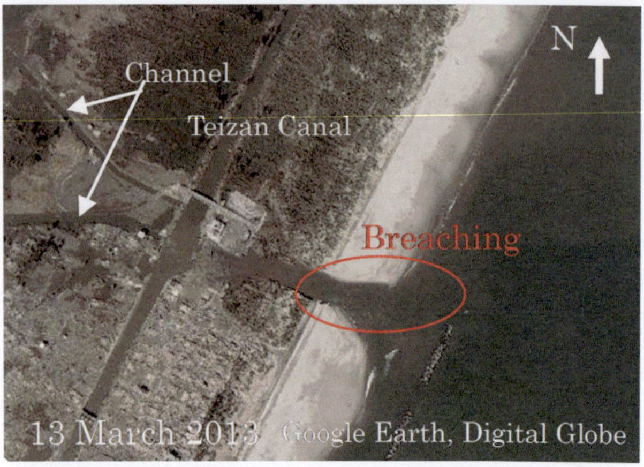

Fig. 9.3 Breaching at Arahama

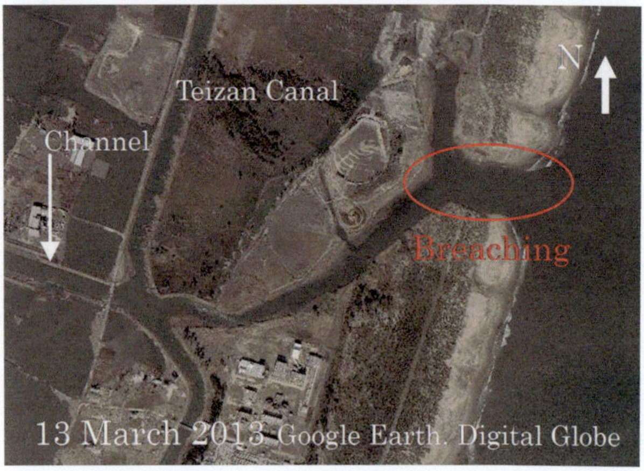

Fig. 9.4 Breaching at Akaiko

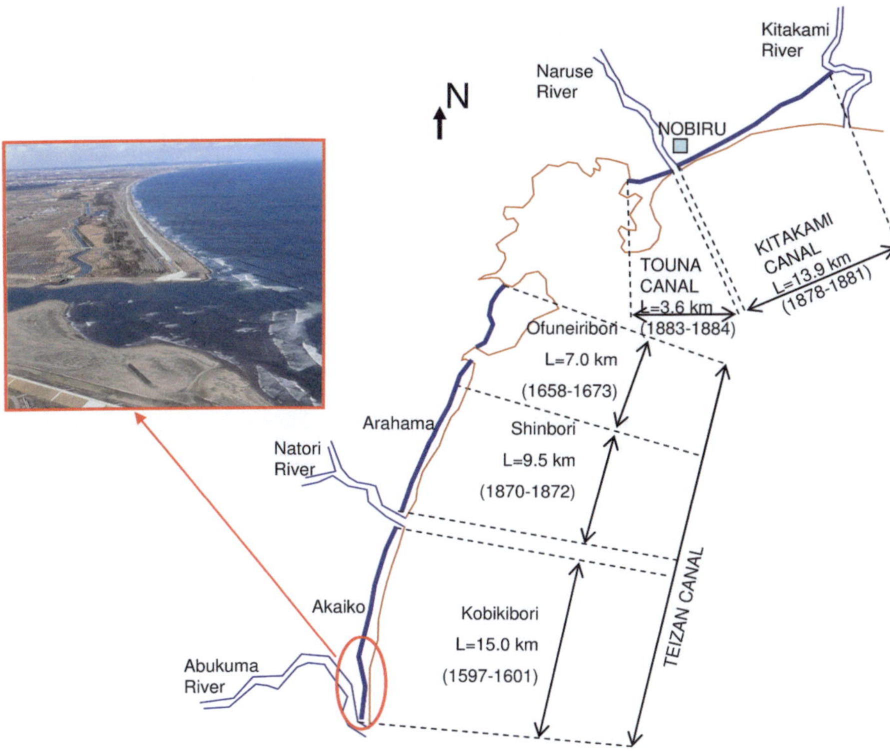

Fig. 9.5 Characteristics and location of the Teizan, Touna, and Kitakami Canal

The Teizan Canal is approximately 25–45 m wide with a maximum depth of about 1.3 m. and has three sections i.e. Kobikibori, Shinbori, and Ofuneiribori which have been constructed in different periods as shown in Fig. 9.5. The Kobikibori canal on the right side of the Natori River was constructed in 1597–1601 stretching from the Natori River to the Abukuma River with a total length of approximately 15 km. Some theories proposed that it was constructed to transport timber for the construction of Sendai Castle as well as the development of Sendai City.

After the construction of the north portion of the Teizan Canal, the Ofuneiribori followed. This 7 km canal was constructed in 1658–1673 to transport goods and rice from the bay to different Sendai areas, bypassing the need to travel long distance along the coastline. The last portion of the Teizan Canal on the left side of the Natori River call the Shinbori, was constructed 270 years after the construction of Kobikibori (1870–1872), with a total length of approximately 9.5 km. The Kobikibori canal was constructed by private merchants supporting trading and transporting timber for the development of Sendai City.

A historical map from 1701 was obtained from Miyagi Prefectural Archives. This map was compared to the satellite image of 2012 as shown in Fig. 9.6. In general, both are comparable although the old map is less accurate, for example the

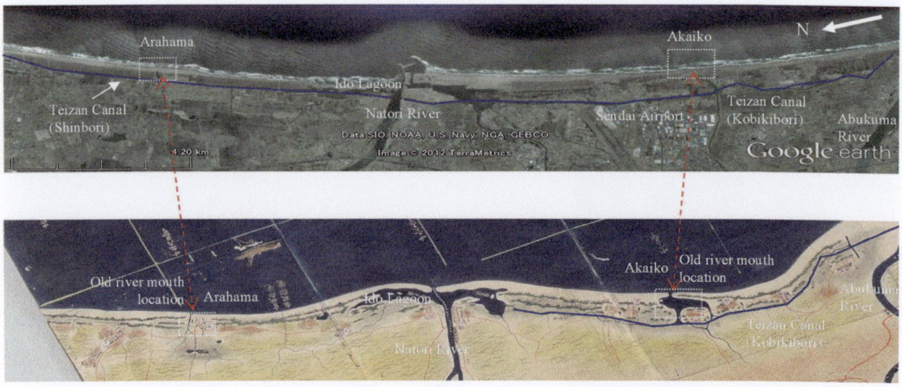

Fig. 9.6 Coastal features comparison between the 1701 and 2012 maps (Source: Google earth and Miyagi Prefectural Archives)

Natori river mouth in the old map was located further on to the seaside than its found in the satellite image. On the other hand, the old map clearly showed the existence of the Teizan Canal-Kobikibori section on the right side of the Natori River as also confirmed in the satellite image. Moreover, the location of the Ido Lagoon at the left side of the Natori River Mouth was also confirmed in both maps. It should be noted here that the portion of the Teizan Canal on the left side of the Natori River (Shinbori) was not yet constructed in 1853. Hence, it was not drawn in the map.

The sandy coastline breaches in both locations, Arahama and Akaiko (Figs. 9.3 and 9.4) occurred where the original river mouths were located based on the comparison of the historical map and the satellite image (Fig. 9.6). The Arahama breaching was also confirmed from an older map drawn in 1853 and obtained from Sendai City Museum (Fig. 9.7). The old river mouths were a part of the old drainage system, at the time the water from the Sendai Plain flowed into the sea through these river mouths closed since the Meiji era (the end of the nineteenth century) as shown in Figs. 9.8 and 9.9. After the closure, the Teizan Canal began collecting and flushing the water off the Sendai plain through the Natori River into the sea. However, as result of the extremely strong 2011 tsunami-induced return flow caused the sandy coast breaching in the old river mouth locations as seen in Figs. 9.3 and 9.4.

9.2.2 Topographical Analysis

The flow direction and water accumulation in the study area were analyzed based on a 5 m resolution DEM data (Udo et al. 2012) as shown in Fig. 9.10. In general, the topographical features at both locations resemble a basin formed towards the breaching location as the exit point. The 3D map shows that the topographical features in Arahama around the breaching point (near the shore) were higher than the

Fig. 9.7 River mouth at Arahama in 1853 (Source: Sendai City Museum)

Fig. 9.8 Arahama (Map at present with route 137)

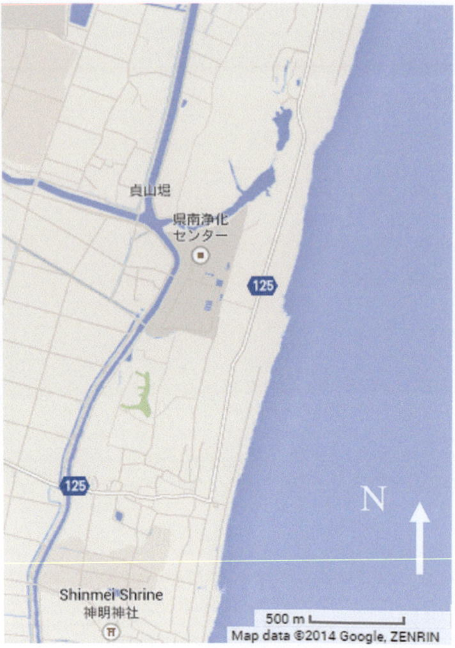

Fig. 9.9 Akaiko (Map at present with Route 125)

inland area due to the existing sand dunes. However, there were traces of past sand dunes, parallel to the present ones showing higher elevation than the surroundings.

The topographical features, along with the canal and local channel in both locations, caused the flow to accumulate towards the old river mouth. Thus, the tsunami-induced return flow gathered at this location leading to the breaching of the sandy coast. For both, Arahama and Akaiko, the topography data were analyzed in more detail around the breaching locations, the topography cross section profiles drawn, and the GIS analysis performed to obtain the water flow direction and accumulation based on the local topographical features using ARCGIS spatial analysis (Flow Direction and Flow Accumulation).

The cross section profiles at Arahama in Fig. 9.11b left, Cross B-B′ and Cross C-C′ shows that the past sand dunes still exist at the south side of the channel. Cross A-A′ and Cross B-B′ show that the Teizan Canal was located higher than its surroundings. Hence, the return flow of the tsunami may not flow into the canal. On the contrary, Cross C-C′ and D-D′ show that the topographical profiles are relatively flat to mild slope towards the local channel. This suggests that the return flow into the local channel, leading to the breaching of the sandy coast around the end of the

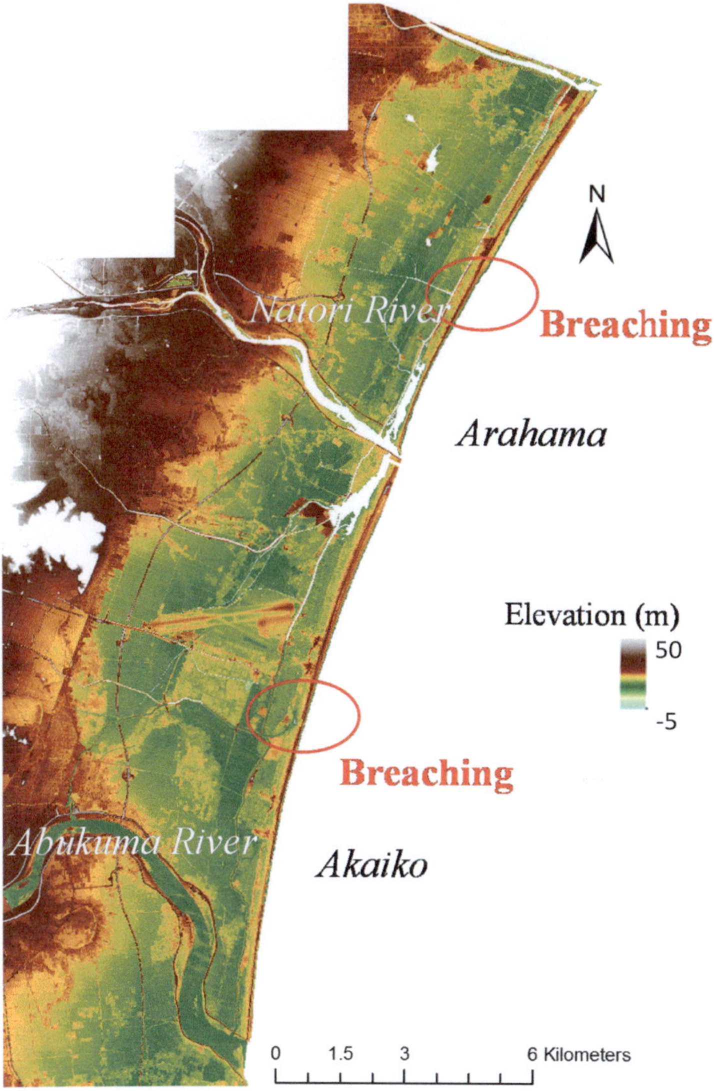

Fig. 9.10 Breaching sites in Arahama and Akaiko. DEM Data (5 m Resolution)

channel. This correlates well with the flow accumulation, shown in Fig. 9.11c left. It also clearly shows that the water flow accumulated towards the old river mouth mainly through the local channel instead of the canal.

The cross section profiles at Akaiko are given in Fig. 9.11b right where Cross A-A′ and B-B′ show that the topography gradient in these sections is relatively mild slope towards the Teizan Canal. Cross C-C′ and D-D′ show that the topography gradient in these sections is also mild slope towards the local channel. Therefore, the tsunami-induced return flow from all directions was likely collected through

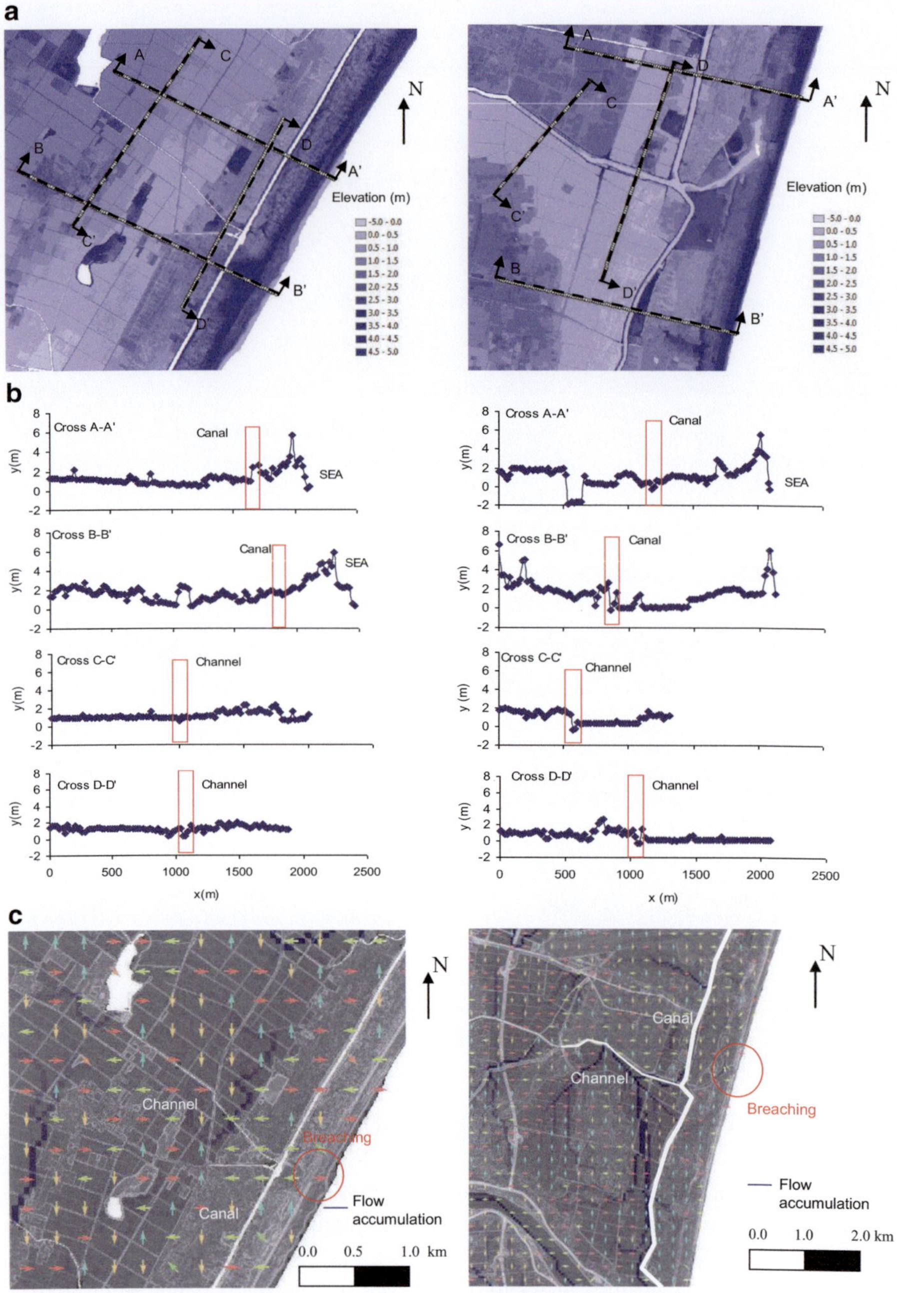

Fig. 9.11 (**a**) (*left*): Countours at Arahama, (*right*): Contours at Akaiko; (**b**) (*left*): Cross section profiles at Arahama, (*right*): Cross section profiles at Akaiko; (**c**) (*left*): GIS flow direction and accumulation. DEM analysis at Arahama, (*right*): GIS flow direction and accumulation. DEM analysis at Akaiko

both, the Teizan Canal and the local channel. This agrees well to the flow accumulation analysis in this location as shown in Fig. 9.11c right. The result shows that the water flow accumulated in both the local channel and the canal. Therefore, the meeting point between both received water flow from all directions (the drained area in this location is also larger than that in Arahama). This situation likely resulted in a more severe breaching than that suffered in Arahama as the estimated breaching width at Akaiko (130 m) was more than twice of that at Arahama (50 m) confirming that the return water flow towards the old river mouth was stronger in Akaiko than in Arahama.

9.3 Conclusions

The present research on the effect of the Teizan Canal and the old river mouths on the tsunami return flooding and sand breaching at Arahama and Akaiko, showed that the latter occurred in both sites even though the river mouths have been closed for many years. By comparing present satellite image with historical maps this finding was confirmed. The breaching is likely to have occurred due to the strong water flow (backwash) induced by the tsunami.

The detailed topography analysis based on a 5 m resolution DEM data showed that the tsunami-induced return flow at Akaiko drained through the Teizan Canal as well as local channel, contrary to the one at Arahama which was mainly collected through the local channel. In addition, the drained area in Akaiko proved to be larger than that of Arahama, making the collected return flow discharge higher in Akaiko causing a more severe breaching at this location.

Topographical features including canals, channels as well as old river mouths have shown to be very important in the process to flush out tsunami return flow and backwash hence reducing the inundation time. In light of this, the present study has furthered shown that these features should also be considered as a part of the tsunami mitigation strategies and plans.

Acknowledgments The authors would like to thank the Sendai Office of River and National Highway, Ministry of Land Infrastructure and Transport for the valuable DEM topographic data. In addition, the video recording taken from the helicopter was provided by Tohoku Regional Development Bureau and the historical maps were provided by Miyagi Prefectural Archives and Sendai City Museum.

References

Adityawan MB, Roh M, Tanaka H, Mano A, Udo K (2012) Investigation of tsunami propagation characteristics in river and on land induced by The Great East Japan Earthquake 2011. J Earthq Tsunami 6:1250033, 22 pages. doi:10.1142/S1793431112500339

Adityawan MB, Dao NX, Tanaka H, Mano A, Udo K (2014) Morphological changes along the Ishinomaki Coast induced by the 2011 Great East Japan Tsunami and the relationship with coastal structures. Coast Eng J 56(3):1450016, 21 pages. doi:10.1142/S0578563414500168

Dao NX, Adityawan MB, Tanaka H, Lin P (2013) Effectiveness of a shore-parallel canal to reduce tsunami impact. In: Proceedings of 34th IAHR World Congress, p 1317

Encyclopedia of the Teizan Canal. http://www.teizanunga.com/Pages/default.aspx. Accessed Feb 2015 (in Japanese)

Mori N, Takahashi T, The 2011 Tohoku Earthquake Tsunami Joint Survey Group (2012) Nationwide post event survey and analysis of the 2011 Tohoku Earthquake Tsunami. Coast Eng J 54(1):1250001, 27 pp. doi:10.1142/S0578563412500015

Shito M, Inuzuka I, Amaya I, Saito H, Kurata J (2014) Numerical simulations and experiments on tsunami for the design of coastal and offshore structures. IHI Eng Rev 46(2):21–25

Suppasri A, Koshimura S, Imai K, Mas E, Gokon H, Muhari A, Imamura F (2012) Damage characteristic and field survey of the 2011 Great East Japan Tsunami in Miyagi Prefecture. Coast Eng J 54(1):1250005, 30 pages. doi:10.1142/S0578563412500052

Tanaka H, Nguyen XT, Umeda M, Hirao R, Pradjoko E, Mano A, Udo K (2012) Coastal and estuarine morphology changes induced the 2011 Great East Japan Earthquake Tsunami. Coast Eng J 54(1):1250010, 25 pages. doi:10.1142/S0578563412500106

Tanaka H, Adityawan MB, Mano A (2014a) Morphology change at the Nanakita River mouth after The Great East Japan Tsunami. Coast Eng 86:14–26. doi:10.1016/j.coastaleng.2014.01.002

Tanaka H, Kayane K, Adityawan MB, Roh M, Farid M (2014b) Study on the relation of river morphology and tsunami propagation in rivers. Ocean Dyn 64(9):1319–1332. doi:10.1007/s10236-014-0749-y

Tappin DR, Evans HM, Jordan CJ, Richmond B, Sugawara D, Goto K (2012) Coastal changes in the Sendai area from the impact of the 2011 tōhoku-oki tsunami: interpretations of time series satellite images, helicopter-borne video footage and field observations. Sediment Geol 282(30):151–174. doi:10.1016/j.sedgeo.2012.09.011

Tsimopoulou V, Jonkman SN, Kolen B, Maaskant B, Mori N, Yasuda T (2012) A multi-layered safety perspective on the tsunami disaster in Tohoku, Japan. In: Comprehensive flood risk management. Taylor & Francis Group, London

Udo K, Sugawara D, Tanaka H, Imai K, Mano A (2012) Impact of the 2011 Tohoku Earthquake and Tsunami on beach morphology along the northern Sendai Coast. Coast Eng J 54(1):1250009, 15 pages. doi:10.1142/S057856341250009X

Usman F, Murakami K, Kurniawan EB (2014) Study on reducing tsunami inundation energy by the modification of topography based on local wisdom. Procedia Environ Sci 20:642–650

Chapter 10
Morphological Characteristics of River Mouths After the 2011 Tohoku Tsunami in Miyagi Prefecture

Min Roh, Yuta Mitobe, and Hitoshi Tanaka

Abstract Coastlines and river mouths along the Tohoku Region in Japan are in the process of steady recovery following the 2011 Tsunami. The affected sandy coastlines and river mouths have undergone more significant changes than the corresponding cliffs, rocky beaches, and hard structures. Analyses of aerial photographs and topographic data related to pre-tsunami and post-tsunami conditions together with the estimated minimum width and positioning changes of river mouths describe how they have changed as well as the differences in their recovery processes. In many cases, there is an indication of stable recovery although the behavioral tendency differs, as in the case of the Natori and Naruse rivers. This study shows that the temporal topographic changes and the relationship with the tidal prism in the mouths of these river mouths differ, resulting in different sediment deposition and restoration processes. Sediment supply is intricately associated with the morphological changes in river mouth morphological changes, which in turn reflect on its recovery. The morphological changes in a river's mouth present practical river management and maintenance problems in ports and harbors; consequently, continuous monitoring is essential.

Keywords Tohoku tsunami • River morphology recovery • Tidal prism

10.1 Introduction

The 2011 Tohoku Earthquake and Tsunami were one of the most devastating natural disasters in history, affecting the society, economy, coastlines, infrastructure, and housing. The tsunami's direct hit and resulting inundation that affected sandy coasts and river mouths was particularly devastating. Four years after this 2011 event, there are signs that many damaged areas of coastlines and river mouths are

---

M. Roh (✉) • Y. Mitobe • H. Tanaka
Department of Civil Engineering, Tohoku University, 6-6-06 Aoba, Sendai, Japan
e-mail: rohmin@kasen.civil.tohoku.ac.jp; yuta.mitobe.c5@tohoku.ac.jp; tanaka@tsunami2.civil.tohoku.ac.jp

V. Santiago-Fandiño et al. (eds.), *Tsunamis and Earthquakes in Coastal Environments*, Coastal Research Library 14, DOI 10.1007/978-3-319-28528-3_10

recovering and returning to their previously existing condition. It is important to learn from the ongoing recovery and restoration process, as they will prove invaluable in river and coastal engineering.

The recovery and restoration process of coasts and river mouths after tsunami hits has been widely studied. Pari et al. (2008) and Choowong et al. (2009) analyzed satellite images and field data to assess the morphological changes and recovery process of beaches and coasts after the 2004 Indian Ocean Tsunami, while Griffin et al. (2013) discussed the responses and recovery process of the coastal environments after the 2004 Tsunami in Aceh (Indonesia), considering factors such as tsunami characteristics, sediment supply, vegetation, and morphology. Bathymetric changes and sediment transportation due to the tsunami impacts were verified through numerical modeling and field survey data by researchers such as Wijetunge (2009), Goto et al. (2011), and Sugawara et al. (2014). Liew et al. (2010) focused on beach recovery and geomorphic processes after high-magnitude low-frequency tsunami events, finding that morphological change and coastal recovery are influenced by the sediment transport pathway and the prevalent longshore current. Udo et al. (2012) assessed morphological changes on the beach along the Sendai Coast due to the 2011 Tohoku Tsunami using remotely sensed data, while Tanaka et al. (2012) described the coastal and estuarine morphological changes in Japan, concluding that the geographical changes in beaches and river mouths are primarily related to the effects of tsunami propagation and land subsidence due to the earthquake, and that sediment supply is one of the most important parameters in the restoration process. Subsequently, Tanaka et al. (2014) proposed a methodology for understanding the morphological changes and recovery of river mouths due to tsunami propagation into rivers that uses water level measurements in rivers and the sea. Although this approach has proven efficient for river mouths, it cannot be used for beaches and other coastal areas.

In this study the morphological changes and recovery processes in river mouths after the 2011 Tohoku Tsunami was assessed by analyzing aerial photographs and bottom topography; estimated minimum width and position changes described how the river mouths formed under pre- and post-tsunami conditions. Topographic data analysis showed the temporal variation of the river mouth bottom as well as the relationship between the inlet cross-sectional area and tidal prism.

10.2 Study Area

The focus of this study was the Natori and Naruse rivers in Sendai and Ishinomaki City, respectively (Fig. 10.1). The national government manages water supply and quality as well as natural extreme disasters in these rivers. The existing jetties at the river mouths were constructed as part of the rivers' shipping and navigation management plan; they were built considering the river flow and coastal sedimentation, which pose a major problem. The mouth of each river exhibits different morphological changes and behavioral processes from the other. The difference may be one

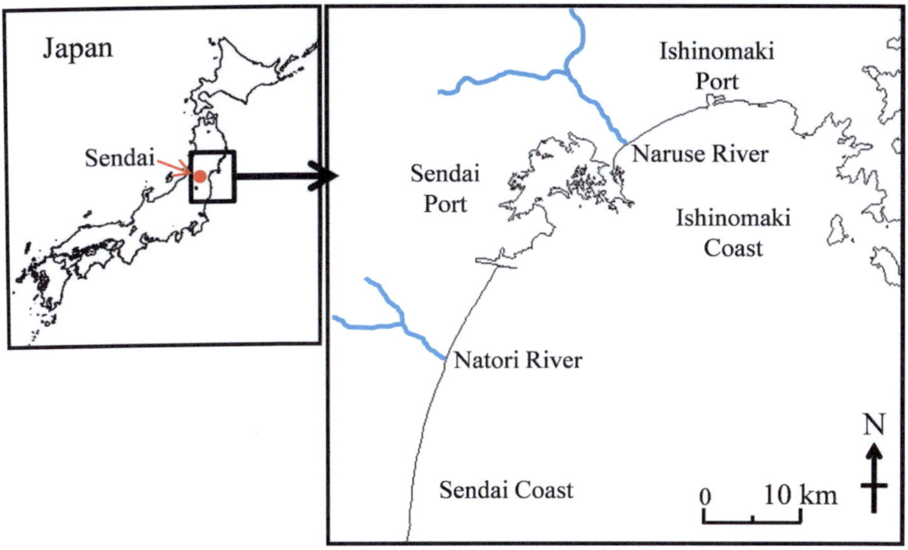

Fig. 10.1 Locations of study area

of the most significant problems in controlling the river mouth system: therefore, investigation of the overall processes based on river mouth morphological changes under pre- and post-tsunami conditions is essential.

10.3 Methodology and Data Collection

Aerial photographs and bottom topography data (provided by the Ministry of Land, Infrastructure, Transport and Tourism -MLIT-) were used to ascertain the recovery process and the morphological changes occurring at the mouth of the rivers. One of the aerial photographs provided by the Geospatial Information Authority of Japan (GIA) was taken on March 13, 2011, immediately after the tsunami struck. High-resolution image processing data have previously been used to analyze the morphology and shoreline changes on beaches and coasts (Fletcher et al. 2003; Ford 2013). In this study, 30 images of the mouth of the Natori River and 24 images of that of the Naruse River during pre- and post-tsunami conditions were utilized to detect river mouth morphological changes, minimum widths, and positions. Bottom topography data near the mouth of the Natori River were measured from 2010 to 2013, whereas the data for the Naruse River were measured from 2011 to 2014. Some cross-sectional data in the Naruse River during pre-tsunami conditions were replaced by measurement data in 2002. Measurement data were used to verify the cross-sectional changes at the mouth of each river due to the impact of the tsunami, with the measurement locations indicating the distance from a given point to the mouth of the river. Standard deviation was then used to determine the temporal changes of the bottom elevation at the given location, and divided into two values depending on whether the bottom elevation under the pre-tsunami condition was included in the calculation. The estimated cross-sectional areas at the river mouth proved useful to assess the relationship with the tidal prism under pre- and post-tsunami conditions.

10.4 Results

10.4.1 River M outh Morphological Changes Under Pre- and Post-Tsunami Conditions

High-resolution photographs were regularly taken every 1 or 2 months prior to and after the earthquake and tsunami, although more recently, they were taken only after intervals of several months.

10.4.1.1 Natori River

Figure 10.2 shows aerial photos of the Natori River taken between March 6, 2011 and November 8, 2014. It is clear from Figs. 10.2a, b that this river mouth was totally destroyed by the tsunami. Following that event, the mouth of the Natori River has been steadily recovering over the years (Figs. 10.2c–f). However, it is important to note that the form of the river mouth has not changed significantly after 2013, although the morphological changes began to stabilize and the river mouth started to be restored, as shown in Figs. 10.2g–i. Nevertheless, the river's present mouth is still different from the one it had prior to the tsunami.

The morphological changes of the mouth of the Natori River between 2011 and 2014, considering pre- and post-tsunami conditions are shown in Fig. 10.3. The river's shoreline was eroded by approximately 300 m near the jetty as a result of the tsunami, as shown in Fig. 10.3a, and thereafter the sand spit at the mouth of the river gradually advanced approximately 100 m behind the jetty (Fig. 10.3b). In Figure 10.3c, the sand spit at the mouth of the river was developing in the upstream direction as well as the narrow width of the river's mouth. As can be seen in Fig. 10.3d, the narrowing of the width of the river's mouth continued unchanged toward the jetty until quite recently. It seems that the river mouth morphology appears as a 500 m straight river channel due to the sediment supply in the direction of flow of the river.

10.4.1.2 Naruse River

The aerial photographs of the mouth of the Naruse River from January 25, 2011 to October 25, 2014 are shown in Fig. 10.4. This river's mouth was also severely affected by the tsunami, as shown in Figs. 10.4a, b. The headland area near the jetty

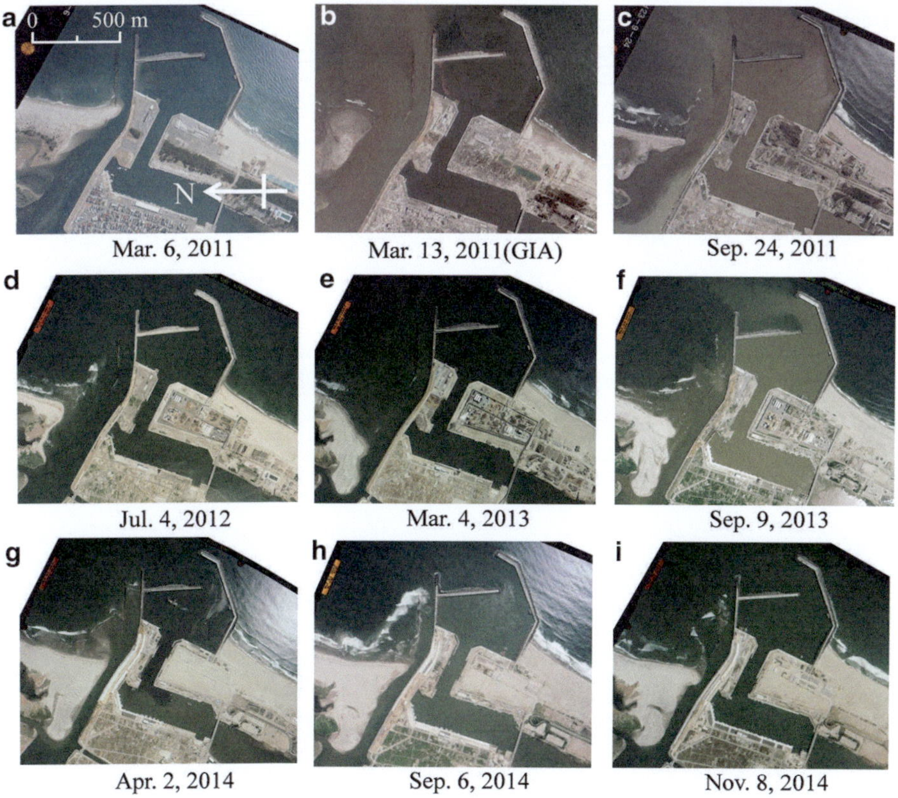

Fig. 10.2 Aerial photographs of the Natori River mouth (Mar. 6, 2011–Nov. 8, 2014)

was washed away as a result of the tsunami's propagation into the river. From Figs. 10.4c–i, it is clear that it has not advanced or recovered as yet, instead, the sediments are deposited in the canal's entrance located on the left-hand side of the river's mouth. This sediment deposition is causing problems with flow control as it is now approaching the port through the canal.

Figure 10.5 shows the temporal morphological changes at the mouth of the Naruse River. In Fig. 10.5a, it is possible to see the section affected by the tsunami, especially the headland area, approximately 300 m long and 250 m wide between the jetty and coastal structure, which was completely washed away. As shown in Figs. 10.5b–d, the headland area has still not recovered to date because the supply of sediment into the mouth of the river is insufficient. However, the morphological changes at the entrance to the canal occurred immediately after the tsunami, and then it gradually advanced toward the mouth of the river. At present, the headland area is located adjacent to the embankment and the small port in the canal.

Considering the above analysis, it is possible to conclude that the respective mouths of the Natori and Naruse rivers developed different morphological characteristics following the tsunami. Moreover, the supply of sediment to the mouth of

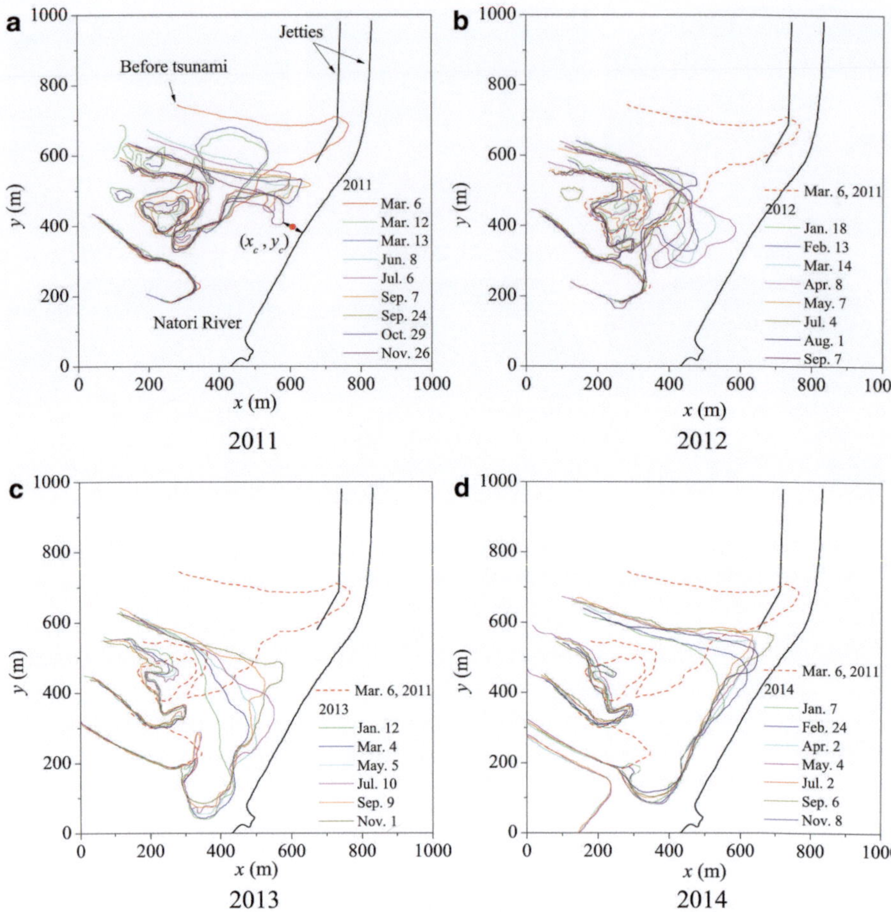

Fig. 10.3 Natori River mouth changes from 2011 to 2014 by analyzing aerial photographs

each river, as well as upstream, may be insufficient, or the morphological changes at the mouth of each river is primarily dominated by the changes at the bottom. In view of this fact, it is necessary to analyze the bottom topography as such an analysis may explain the overall process of morphological changes at the mouth of the rivers.

10.4.2 Estimations of Minimum River Mouth Widths and Position Changes

The changes at the mouths of the Natori and Naruse rivers were obtained using the minimum widths of their mouths and also the changes in their positions, which can be considered as the balance positions between the original and current flow of a river. The positional changes indicate the dominant direction of the morphological

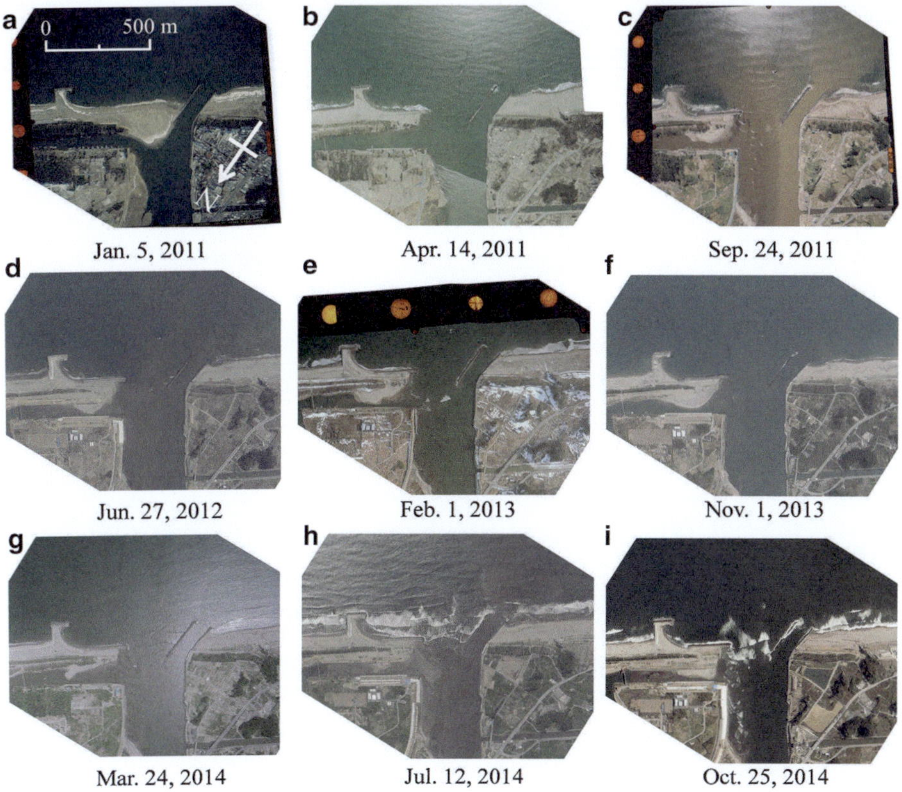

Fig. 10.4 Aerial photographs of the Naruse River mouth (Jan. 5, 2011–Oct. 25, 2014)

changes in the river mouth in the longshore direction (x_c) and in the offshore-onshore direction (y_c). It can be estimated in the midpoint based on the straight line on the minimum river mouth width, as shown in Fig. 10.3a and Fig. 10.4a, respectively.

The minimum river mouth width (B_{min}) for the Natori River is shown in Fig. 10.6. The width of the river mouth increased to approximately 130 m as a result of the tsunami. The minimum river mouth width experienced fluctuations until the end of 2012, and thereafter, it stabilized at about 50 m from the beginning of 2013. The minimum river mouth width is approximately one-half of the jetty's width. Figure 10.7 shows the positional changes of the minimum river mouth width under pre- and post-tsunami conditions. It can be seen that the river mouth has been restored mainly on the riverside in the offshore-onshore direction, as opposed to the longshore direction.

The estimated minimum river mouth width for the Naruse River is plotted in Fig. 10.8. The figure shows that immediately following the tsunami, the width virtually doubled compared to the pre-tsunami conditions, but thereafter, within the next 3 months after the tsunami, it significantly decreased by approximately 75 m, although during that time, the minimum river mouth area was restored to 500 m behind the jetty. Since the middle of 2011, the width of the river mouth has again rapidly increased, and currently, it is approximately 220–230 m. The river mouth

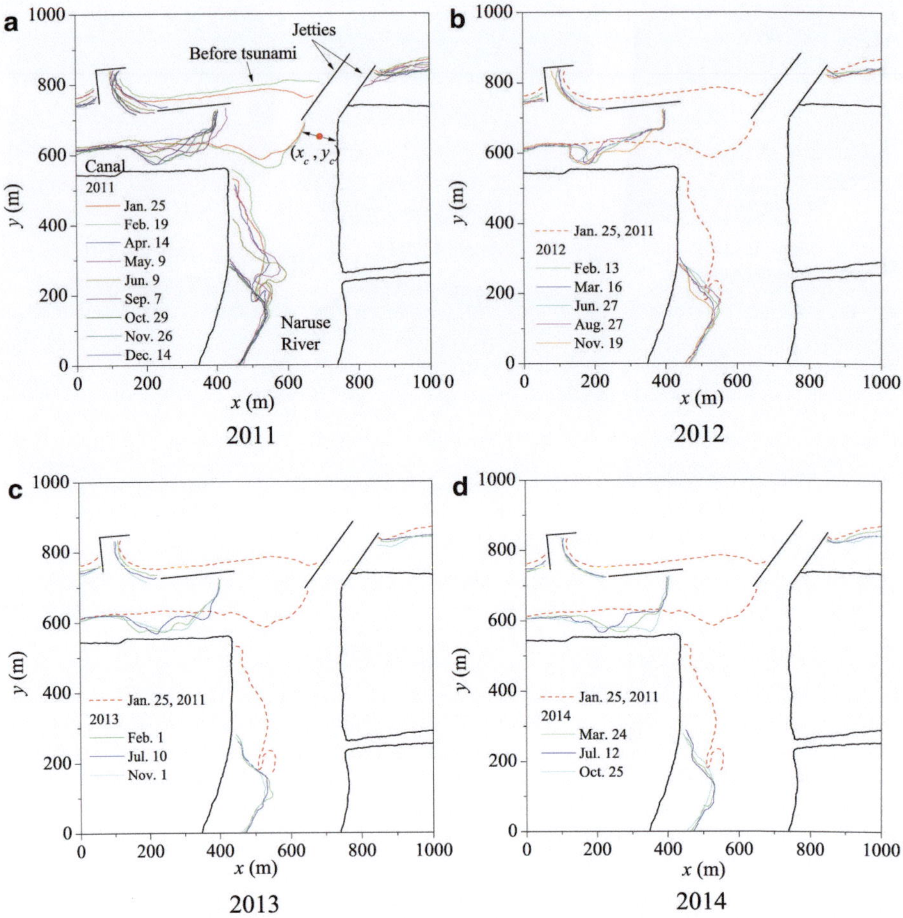

Fig. 10.5 Naruse River mouth changes from 2011 to 2014 by analyzing aerial photographs

has maintained a constant minimum width since the middle of 2011. The current minimum width is virtually three times that of the jetty. Figure 10.9 shows the changes in the position of the minimum river mouth width of the Naruse River.

The mouth of the Naruse River also appeared to be moving in an offshore-onshore direction, especially along the riverside. The directions and the levels according to the positional changes of the minimum river mouth width are clearly depicted in the results of the linear regression.

10.4.3 Temporal Changes of River Mouth Bottom Topography

Standard deviation (SD) was used to verify the changes in topography under pre- and post-tsunami conditions. The differences in the SDs along the cross-section data varied according to whether the pre-tsunami data were included or not. The

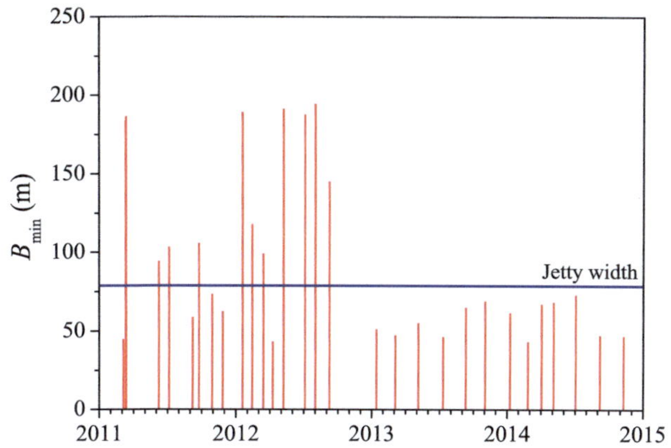

Fig. 10.6 Estimated minimum river mouth width in the Natori River

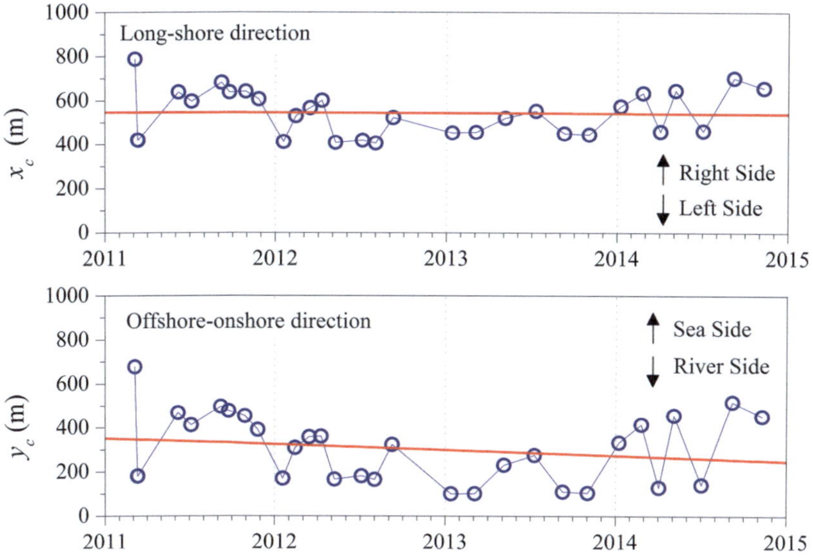

Fig. 10.7 Position changes of the minimum river mouth width in the Natori River

temporal variations of the bottom topography at the four cross-sections of the Natori River are depicted in Fig. 10.10. The maximum changes in the bottom topography were estimated at –0.6 km from the mouth of the river toward the Pacific Ocean. Furthermore, the amount of erosion due to the tsunami, and sediment deposition could be identified by analyzing the bottom topography. In this case, a severe case of erosion was discovered at the mouth of the river, around the jetty; thereafter, its bottom conditions showed deposited sediment in the river channel. At 0.6 km from the mouth of the river, the deposits were approximately 3 m deep and sediment deposition resulted in the bottom becoming flat. The high SDs in both cases were

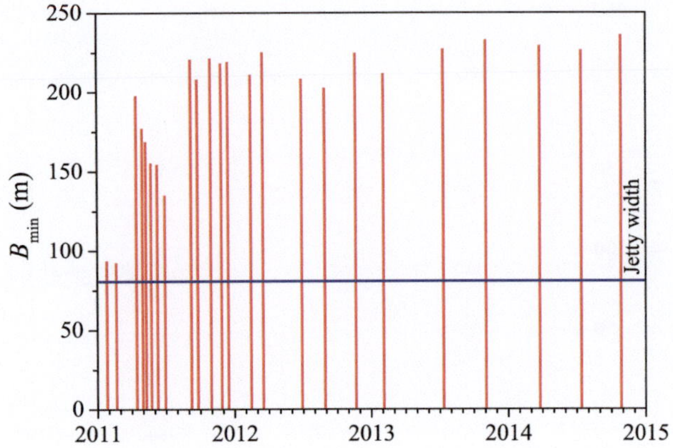

Fig. 10.8 Estimated minimum river mouth width in the Naruse River

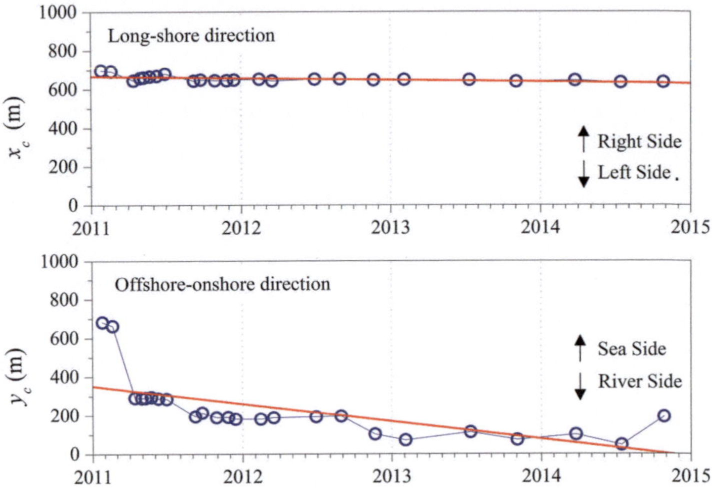

Fig. 10.9 Position changes of the minimum river mouth width in the Naruse River

estimated in the range from 400 to 500 m, whereas the sediment supply was insufficient up to 350 m. Therefore, the SD for pre- and post-tsunami conditions is greater than that for which only post-tsunami conditions are considered. Locations –0.4 and –0.2 km from the river mouth exhibited similar SDs in both cases depending on the bottom elevation under pre-tsunami conditions. At the mouth of the river (0.0 km), large volumes of sediment were supplied over a range of 250–450 m since August 2012. The highest SD was calculated for the large changes in the bottom elevation.

Detailed topography data of the Naruse River can be used to obtain the vertical temporal changes at the bottom of the mouth of the river. As seen in Fig. 10.11, the

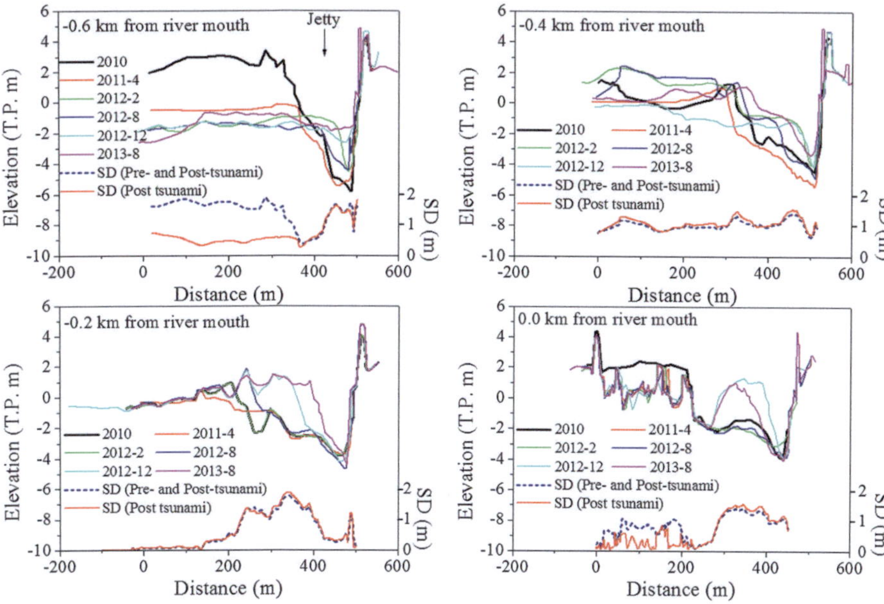

Fig. 10.10 Temporal changes of the bottom cross-section of the Natori River mouth

bottom elevations and the SDs are divided into the cross-section located at the mouth of the river to 1.0 km thence. In cases from the mouth of the river (0.0 km) to 0.16 km thence, a sand bar covered a distance ranging from 100 to 300 m before the tsunami. However, this was totally destroyed by the tsunami; the SD comprising both the pre- and post-tsunami data was higher than that comprising the post-tsunami data only. The upstream section of the river and the surrounding natural environment were not severely affected by the tsunami. The SDs were estimated at low level for pre- and post-tsunami conditions along the cross-section.

10.4.4 Comparison With Tidal Prism and the Inlet Cross-Sectional Area

The inlet cross-sectional areas (A_c) at the mouth of the Natori and Naruse rivers were obtained from the bottom topography data. The mean cross-sectional and minimum cross-sectional areas were determined for the entire river mouth at the same time. As seen in Fig. 10.12, the estimated temporal variations of the inlet cross-sectional areas are divided for the Natori River and the Naruse River. The inlet area in the Naruse River is smaller than that of the Natori River before the tsunami. Subsequently, the inlet areas of both rivers increased significantly to approximately 800 m^2 as a result of tsunami propagation into the rivers. Then, the cross-sectional areas gradually decreased over the years. However, the recovery time for each river

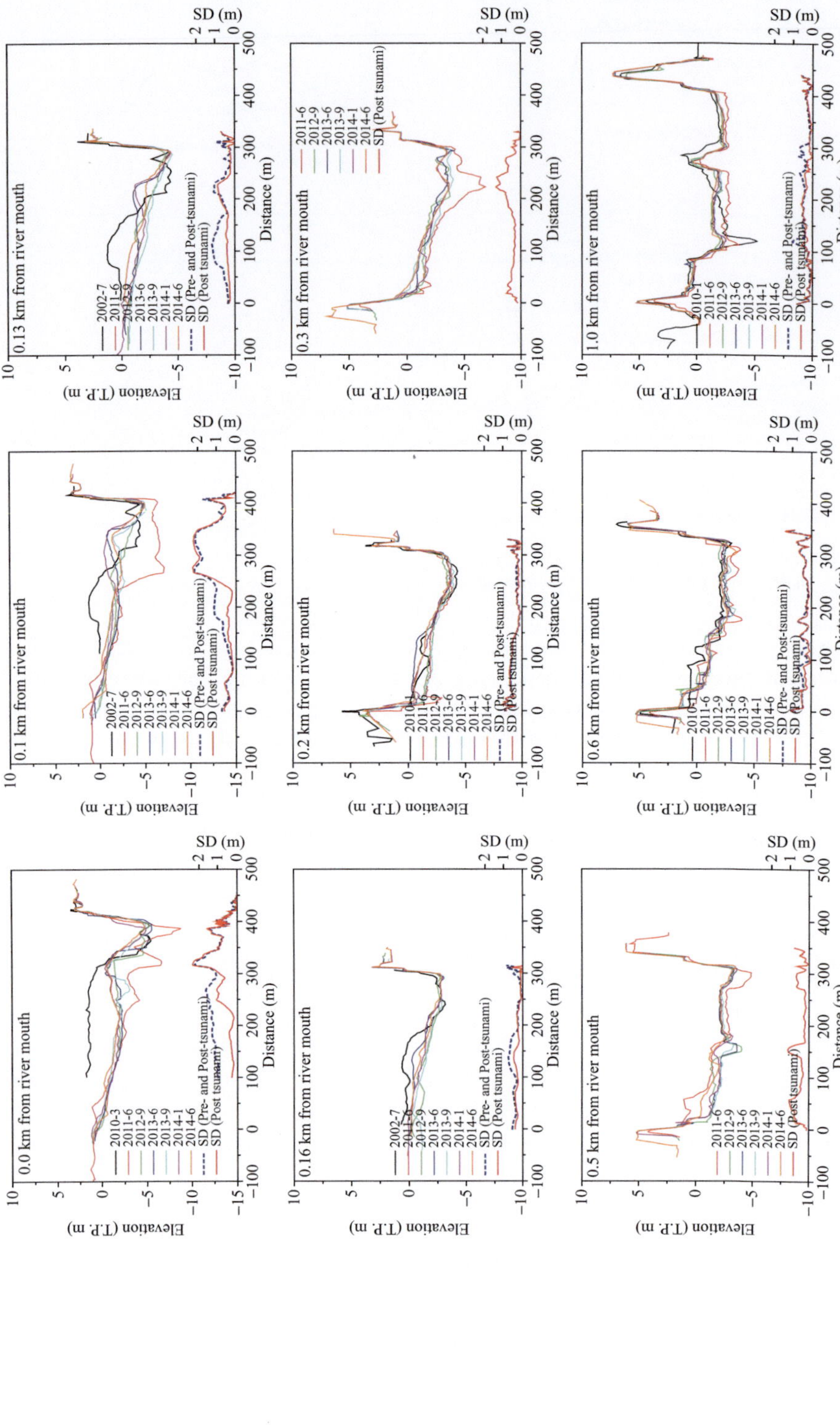

Fig. 10.11 Temporal changes of the bottom cross-section of the Naruse River mouth

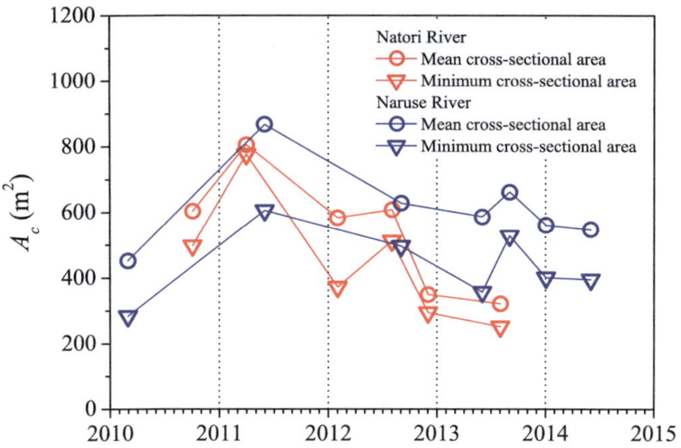

Fig. 10.12 Comparison the inlet cross-sectional areas (A_c) of the Natori River and the Naruse River

is different; the Natori River has recovered in a relatively short time, whereas the inlet area in the Naruse River is still larger than that of the Natori River. This difference may be caused by sediment transportation and sediment supply into the river. Consequently, several practical problems have appeared. For example, during heavy flooding, the mouth of the Natori River can become inundated and overflow as a result of the large amount of sediment deposition reforming the mouth of the river. At the mouth of the Naruse River, the sediment is transported to the canal entrance, not to the river. The arrangement results in difficulty approaching the port through the canal. On the other hand, control of the flow of the river may be easier than in the case of the Natori River during floods. The present situation in both rivers due to the morphological changes at the mouth of the river has become a serious and important problem in society.

The tidal prism to evaluate changes in the volume of water in an estuary is strongly related to the inlet cross-sectional area, tidal level, and river surface area up to the limit of the tidal influence in the upstream area of a river (O'Brien 1969; Jarrett 1976; Watanabe et al. 2003). In this study, the spring tide level and the river surface area up to the end of the tidal effect were used to calculate the tidal prism. The limit of tidal effect in the river was estimated based on the mean sea level on the longitudinal river bottom data. Figure 10.13 shows the relationship between the inlet cross-sectional area (A_c) and the tidal prism (P) in the Natori and Naruse rivers. The relationship at the mouth of each river may be different because of the difference in their recovery and restoration processes. This result can be used to deal with the practical problems at the mouth of both the rivers. The problem is that a large amount of sediment causes river management and flow control related difficulties in the Natori River, whereas the Naruse River may have relatively small amounts of sediment deposition. This makes it difficult to control the river mouth system. Therefore, continual and detailed monitoring and field surveying are required to understand the natural phenomena in these river systems. The morphological

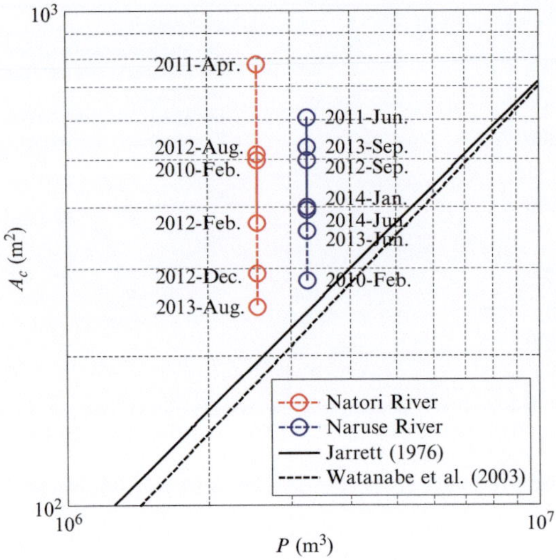

Fig. 10.13 Relationship between the inlet cross-sectional areas (A_c) and the tidal prism (P)

changes and recovery processes at the mouth of the rivers should be studied and discussed more actively including various perspectives in the future.

10.5 Conclusions

In this study, the morphological changes in the mouth of rivers due to the 2011 Tohoku Tsunami were assessed by analyzing related aerial photographs and bottom topography data. Further, overall recovery of the Natori and Naruse rivers was successfully evaluated.

The estimated minimum widths and positional changes in the mouths of the rivers were used to evaluate the morphological changes under pre- and post-tsunami conditions. These results showed that, morphologically, the two river mouths have been in a stable state. Reformation of each river's mouth had advanced from the riverside direction after the tsunami in both rivers. However, the tendency in the recovery and morphological characteristics indicate different behaviors at the mouth of the Natori River and that of the Naruse River, respectively. The morphology of the Natori River mouth is recovering to the state it was in before the 2011 Tohoku Tsunami, whereas, the present morphology of the mouth of the Naruse River began a new stage in the recovery and reformation process. The different processes at the river mouth areas confirmed the vertical cross-sectional changes and the relationship between the inlet cross-sectional area and the tidal prism obtained from the bottom topography data. The morphological changes and the recovery process are found to have been affected by the sediment supply into the river channel.

The research results obtained show the overall process of morphological changes in the mouth of the rivers under pre- and post-tsunami. The analysis method can be applied to the extended areas damaged by the 2011 Tohoku Tsunami. However, the analysis method depends heavily on the availability of monitoring and measurement data. It is necessary to obtain continuous monitoring and observation data such as high-resolution image data and topography data. These studies will help to facilitate understanding of the general process and mechanism associated with river mouth morphological changes and recovery.

Acknowledgments We sincerely thank Kitakamigawa-Karyu (Downstream region of Kitakami River) River Office in Tohoku Regional Bureau of Ministry of Land, Infrastructure, Transport and Tourism (MLIT) for the valuable topography data and aerial photographs used in this study. We are also grateful for financial support in the form of a Grant-in-Aid for Scientific Research from JSPS (No. 2301367) and the River Environmental Fund (REF) from the Foundation of River and Watershed Environmental Management (FOREM). The first author is a JSPS Postdoctoral Fellow (No. 25-03364).

References

Choowong M, Phantuwongraj S, Charoentitirat T, Chutakositkanon V, Yumuang S, Charusiri P (2009) Beach recovery after 2004 Indian Ocean tsunami from Phangnga, Thailand. Geomorphology 104(3–4):134–142

Fletcher C, Rooney J, Barbee M, Lim SC, Richmond B (2003) Mapping shoreline change using digital orthophotogrammetry on Maui, Hawaii. J Coast Res 106–124

Ford M (2013) Shoreline changes interpreted from multi-temporal aerial photographs and high resolution satellite images: Wotje Atoll, Marshall Islands. Remote Sens Environ 135:130–140

Goto K, Takahashi J, Oie T, Imamura F (2011) Remarkable bathymetric change in the nearshore zone by the 2004 Indian Ocean tsunami: Kirinda harbor, Sri Lanka. Geomorphology 127:107–116

Griffin C, Ellis D, Beavis S, Zoleta-Nantes D (2013) Coastal resources, livelihoods and the 2004 Indian Ocean tsunami in Aceh, Indonesia. Ocean Coast Manag 71:176–186

Jarrett JT (1976) Tidal prism – inlet area relationships. U. S. Army Coastal Engineering Research Center and U. S. Army Engineers Waterway Experiment Station G.I.T.I. Report 3

Liew SC, Gupta A, Wong PP, Kwoh LK (2010) Recovery from a large tsunami mapped over time: the Aceh coast, Sumatra. Geomorphology 144(4):520–529

O'Brien MP (1969) Equilibrium flow areas on inlets on sandy coasts. J Waterw Harb Div ASCE 95(WW1):43–52

Pari Y, Ramana Murthy MV, Jaya Kumar S, Subramanian BR, Ramachandran S (2008) Morphological changes at Vellar estuary, India – impact of the December 2004 tsunami. J Environ Manag 89:45–57

Sugawara D, Takahashi T, Imamura F (2014) Sediment transport due to the 2011 Tohoku-oki tsunami at Sendai: results from numerical modeling. Mar Geol 358:18–37

Tanaka H, Tinh NX, Umeda M, Hirao R, Pradjoko E, Mano A, Udo K (2012) Coastal and estuarine morphology changes induced by the 2011 Great East Japan Earthquake Tsunami. Coast Eng J 54(1):1250010

Tanaka H, Adityawan MB, Mano A (2014) Morphological changes at the Nanakita River mouth after the Great East Japan Tsunami of 2011. Coast Eng 86:14–26

Udo K, Sugawara D, Tanaka H, Imai K, Mano A (2012) Impact of the 2011 Tohoku Earthquake and Tsunami on beach morphology along the northern Sendai Coast. Coast Eng J 54(1):1250009

Watanabe K, Tanaka H, Tsukiyama T (2003) Morphological change at the Natori River mouth caused by reduction of tidal compartment. Proceedings of XXX IAHR Congress, Theme A 547–553

Wijetunge JJ (2009) Field measurements and numerical simulations of the 2004 tsunami impact on the east coast of Sri Lanka. Pure Appl Geophys 166:593–622

Chapter 11
Post-Tsunami Lagoon Morphology Restoration Sendai; Japan

Vo Cong Hoang, Hitoshi Tanaka, and Yuta Mitobe

Abstract The 2011 Great East Japan Earthquake and Tsunami caused significant changes of morphology of Gamo Lagoon which is located in the northern part of Sendai Coast, Miyagi Prefecture. The investigation on the morphology changes and its recovery process after the tsunami are discussed. Frequently captured aerial photographs are effectively utilized to analyze detailed morphological changes in response to wave action after the tsunami. The lagoon area was greatly changed due to the 2011 tsunami, and a gradual process of river mouth changes and restoration were observed using aerial photographs. As compared with the shoreline position before the tsunami a 40 m retreat of the shoreline can be confirmed even after 3 years from the tsunami. Meanwhile, the water area shows distinct reduction by 40 % as compared with the pre-tsunami situation. Gamo Lagoon is well known not only having important roles in the coastal processes but also valuable brackish water environment, hence further investigation is also largely required to better understand the future state of the devastated coastal environment by the tsunami waves.

Keywords Gamo Lagoon • The 2011 tsunami • Sandy barrier • Breaching • Erosion • Recovery

11.1 Introduction

The Great East Japan Earthquake Tsunami occurred on March 11, 2011 in the northern part of Japan. Besides severe damages of infrastructures, it also caused significant morphological changes in the coastal area. As tsunamis move sediments

---

V.C. Hoang
Department of Civil Engineering, Tohoku University, 6-6-06 Aoba, Sendai, Japan

Thuyloi University - Southern Campus, 02 Truong Sa, Binh Thanh, Ho Chi Minh, Vietnam
e-mail: vo.cong.hoang.t3@dc.tohoku.ac.jp; hoangvc@tlu.edu.vn

H. Tanaka • Y. Mitobe (✉)
Department of Civil Engineering, Tohoku University, 6-6-06 Aoba, Sendai, Japan
e-mail: hitoshi.tanaka.b7@tohoku.ac.jp; yuta.mitobe.c5@tohoku.ac.jp

V. Santiago-Fandiño et al. (eds.), *Tsunamis and Earthquakes in Coastal Environments*, Coastal Research Library 14, DOI 10.1007/978-3-319-28528-3_11

around, wave height of historical events can be estimated from sediment deposition layers (Goto et al. 2014; Takashimizu et al. 2012; Minoura et al. 2001).

Tanaka et al. (2012) reported the changes and subsequent recovery process of the coastal and estuarine morphology in Miyagi Prefecture (Japan) after the tsunami. Several kinds of morphological changes have been pointed out including beach and sandy barrier erosion, breaching of sandy coast, etc. The erosion of sandy barriers located in front of lagoons leaving them opened and exposed to the sea is one of the most common changes found in the morphology of the Sendai Coast in this prefecture. Gamo Lagoon, located to the left side of the Nanakita River mouth in the northern part of Sendai Coast, is an example as a 2 km long sandy barrier in front of this lagoon was almost totally eroded leaving only a few sandy islands standing. Moreover, deposited sediments covered most of the surface of the lagoon behind the barrier.

Large numbers of experts have focused their research on the significant changes and recovery process of the coastal and estuarine morphology after tsunamis. Since the 2004 Indian Ocean Tsunami, Ali and Narayana (2015) discussed the short-term morphological response and shoreline changes as well as the recovery at Trinkat Island, Andaman and Nicobar, India. And prior to that, Liew et al. (2010) and Choowong et al. (2009) presented the damage and recovery process of beach and estuarine morphology of affected areas in Indonesia and Thailand, respectively. These studies utilized yearly and medium resolution satellite images after the tsunami to discuss the recovery of morphology of the coastline.

The Sendai Coast and surrounding areas have also been the focus of a number of studies on the morphological damage and recovery after the 2011 Great East Japan Earthquake Tsunami, they include those by Hoang et al. (2015) showing the recovery of the tsunami-induced concave shoreline at Gamo Lagoon area; Tanaka et al. (2014) investigating the morphological evolution of the Nanakita River mouth closure based on water level data taken from the inner and the outer sections of the river's mouth. Moreover, Tappin et al. (2012) and Udo et al. (2012) discussed the coastal morphology changes on Sendai Plain and its subsequent recovery based on aerial photographs and bathymetry data sets.

In the case of Gamo Lagoon, studies indicate that the recovery of the shoreline position is nearly similar to that before the tsunami occurred. However, the full recovery of the morphology of the lagoon, particularly with regard to the shoreline position and water area is still missing.

As Gamo Lagoon is a valuable brackish water environment, detailed investigations on water quality, ecology and the environment have already been undertaken. Shin and Nishimura (2013) estimated food sources and utilization in the bivalve *Nuttallia olivacea* before and after the tsunami took place. Sakamaki et al. (2006) proposed mechanism of nutrient flux variation in this lagoon. Moreover, analysis on the relationship between shallow exposed inshore nursery grounds or estuarine nursery grounds and population of stone flounders (*K. bicoloratus*) have proven that the estuarine environment including that of the lagoon, play an important role as nursery grounds for juveniles of stone flounders (Malloy et al. 1996; Omori et al. 1976; Yamashita et al. 2000).

The morphological recovery of the shore and the lagoon's surface area, which were serve damaged by the tsunami, is most important not only for coastal morphology management, but also for ecosystem. Furthermore, findings on that recovery process will provide useful information for the future similar disasters preparation and recovery planning.

In light of the above, the continuation of the studies on the recovery of Gamo Lagoon morphology becomes imperative. This could be done through the analysis of high resolution and frequent aerial photographs.

11.2 Study Area and Data Collection

Gamo Lagoon is located on the northern part of Sendai Coast, Miyagi Prefecture, Japan (Fig. 11.1). It is connected to Nanakita River located on the south side of this lagoon. This river has a length of 45 km with a basin area of 229.1 km^2, and the average annual discharge equals 10 m^3/s.

The lagoon, which used to be the former river mouth channel of Nanakita River, has rich biodiversity, and it is also an important destination place for migration birds during winter time.

Because the direction of longshore sediment transport is from south to north, the sand spit developed toward the north. The location of former river mouth was about 2 km northeast of the current river mouth. Since the start of the construction of Sendai Port in 1967, the river mouth was fixed and a new river mouth was artificially opened at the current location, and in order to stabilize its morphology a jetty was constructed on its left side. The river channel running between both mouths formed Gamo Lagoon.

The high resolution (nearly 0.5 m) aerial photographs of this area have regularly been taken by airplane with a frequency of every one or two months since 1990. In the present study to obtain more detailed changes on the morphology's features induced by the tsunami, aerial photographs taken by the Geospatial Information

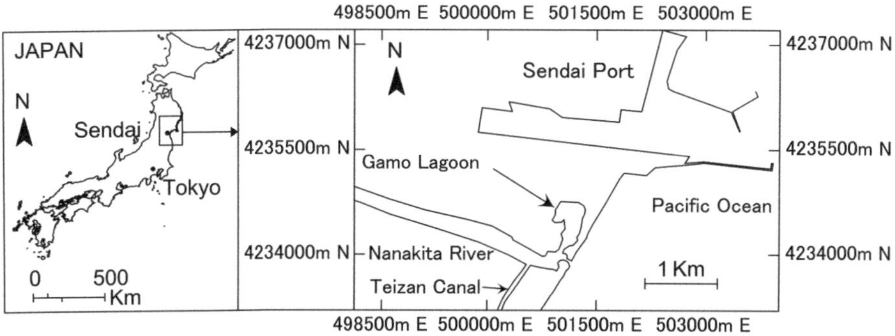

Fig. 11.1 Location map of study area

Authority of Japan (GSI) on March 12, 2011 (1 day after the tsunami), have also been utilized. In addition, aerial photographs from Google Earth to explain in more details the subsequent recovery process after the tsunami have been used. The definition of shoreline and its position have been discussed by Elizabeth and Ian (2005). Moore (2000) has given an overview of coastal mapping techniques which have been developed over the last 27 years. In this study, the wet/dry line is selected to represent the shoreline position. The different colors of pixels between the wet and the dry sides in an aerial photograph are taken as the reference for the shoreline position extraction. Prior to the shoreline position extraction, all of the raw aerial photographs are geo-referenced to the World Geodetic System (WGS-84). A line with a 210° clockwise north direction is taken as a baseline, and the shoreline positions are extracted in every 20 m along this baseline from the aerial photographs. In order to eliminate the effect of tide, detected shoreline positions have been corrected with astronomical tide level in Sendai Port calculated by Japan Meteorological Agency and average beach slope of 0.11 (Hoang et al. 2014). It is noted that shoreline positions extracted from GSI and Google Earth images are not corrected with tidal level due to lack of exact time of capturing. Waterline, which is the boundary between sandy barrier and the water area of the lagoon, are also extracted in the same way with the shoreline. Tidal correction has not been applied to this measurement. More details on the uncertainty assessment of image analysis have been discussed by Pradjoko and Tanaka (2010).

11.3 Morphological Changes of Gamo Lagoon

11.3.1 Morphological Changes Before the Great East Japan Earthquake and Tsunami

Before the tsunami, one of the most significant morphological changes regarding to this lagoon is the sediment intrusion induced by the wave overtopping during the high wave period (Tinh and Tanaka, 2011), and this resulted in the lagoon becoming shallower. Concerned by the changes in the lagoon's water quality and its natural environment, a committee was established in 2005 to look at its restoration. Hence a number of actions have been implemented to improve the water quality. A channel was dug to connect the area on the left side of the lagoon to the lagoon entrance which is near the Nanakita River mouth. A concrete structure to prevent wave overtopping was also constructed.

Figure 11.2a shows the river mouth morphology before the 2011 tsunami. As the predominant direction of longshore sediment transport is from south to north, the development of sand spit on the right side of the Nanakita River mouth is remarkably observed. A jetty was constructed on the left side of the river mouth, and as a result, morphology in this side has been stabilized and fixed. Structure for preventing wave overtopping can be seen on the left hand side of this photograph. More details on morphological changes of Gamo Lagoon in the period before the tsunami can be found from Tanaka et al. (2002).

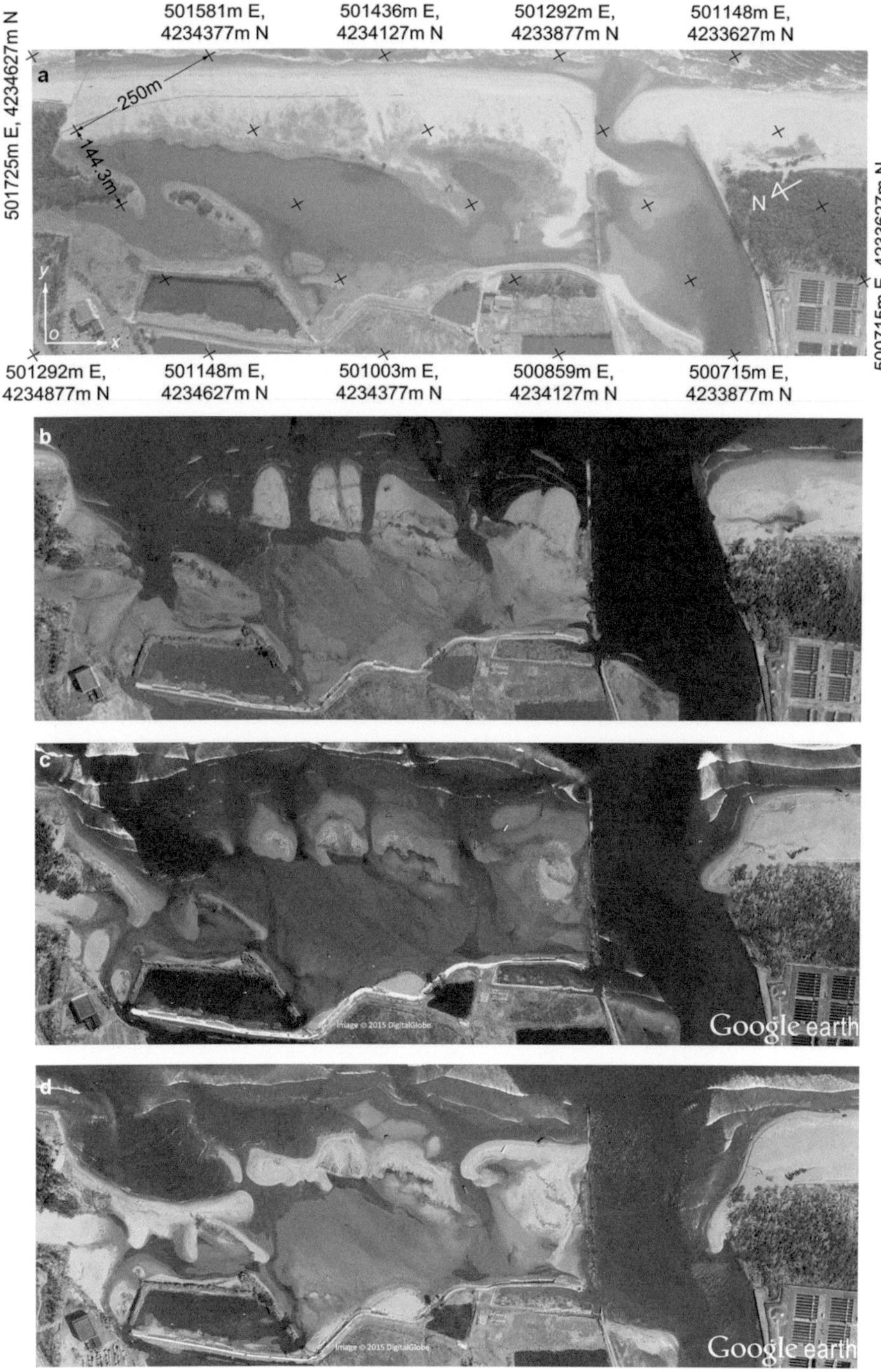

Fig. 11.2 Morphological changes and recovery at Gamo Lagoon. (**a**) March 6, 2011, (**b**) March 12, 2011, (**c**) March 24, 2011, (**d**) April 6, 2011 (*above*), (**e**) May 3, 2011, (**f**) June 8, 2011, (**g**) July 6, 2011, (**h**) September 7, 2011, (**i**) September 24, 2011, (**j**) March 14, 2012, (**k**) April 8, 2012, (**l**) July 4, 2012, (**m**) September 7, 2012, (**n**) January 12, 2013, (**o**) March 4, 2013, (**p**) May 5, 2013, , (**q**) September 9, 2013, (**r**) February 24, 2014 (*below*). The read arrow indicates a marked change in the morphology

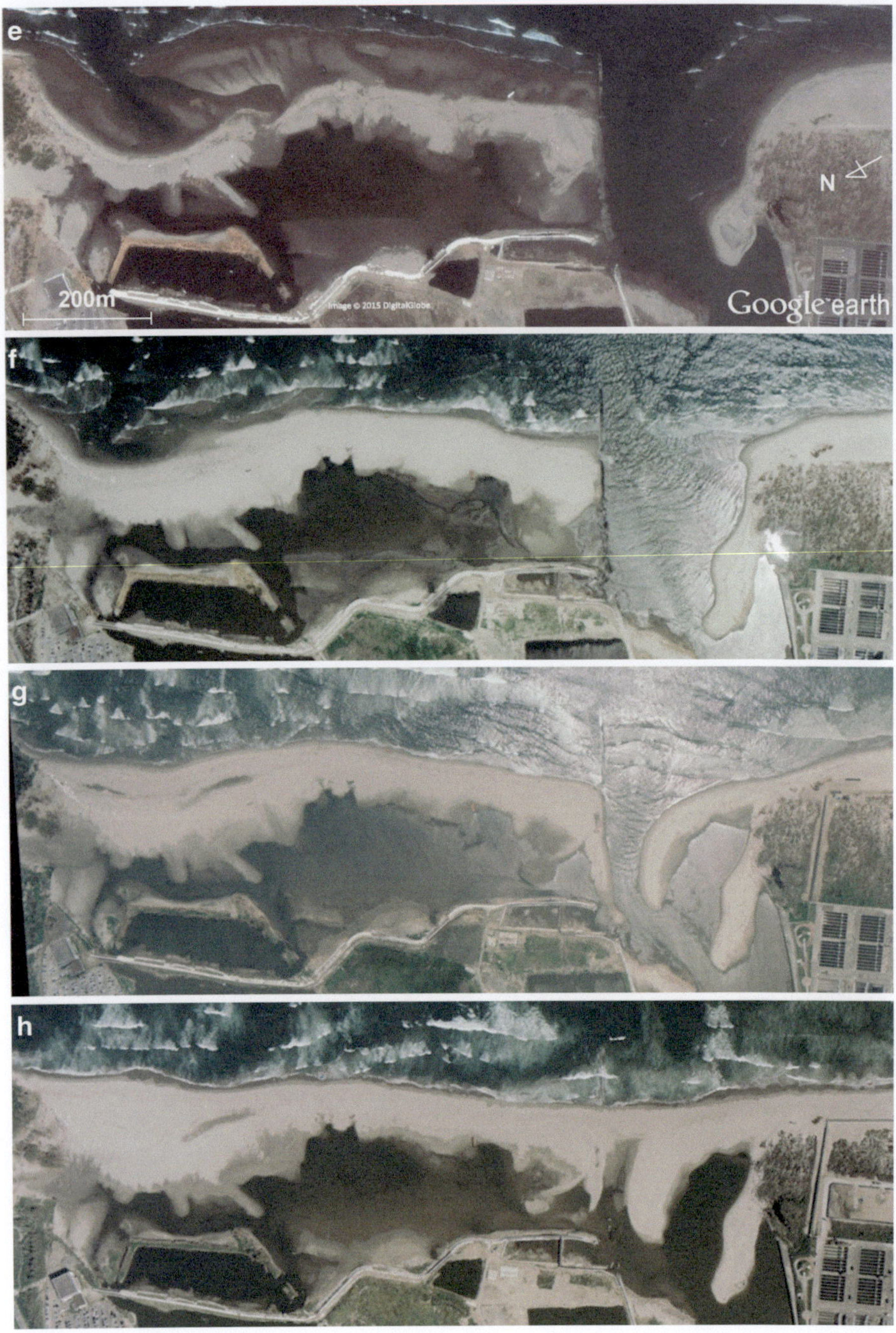

Fig. 11.2 (continued)

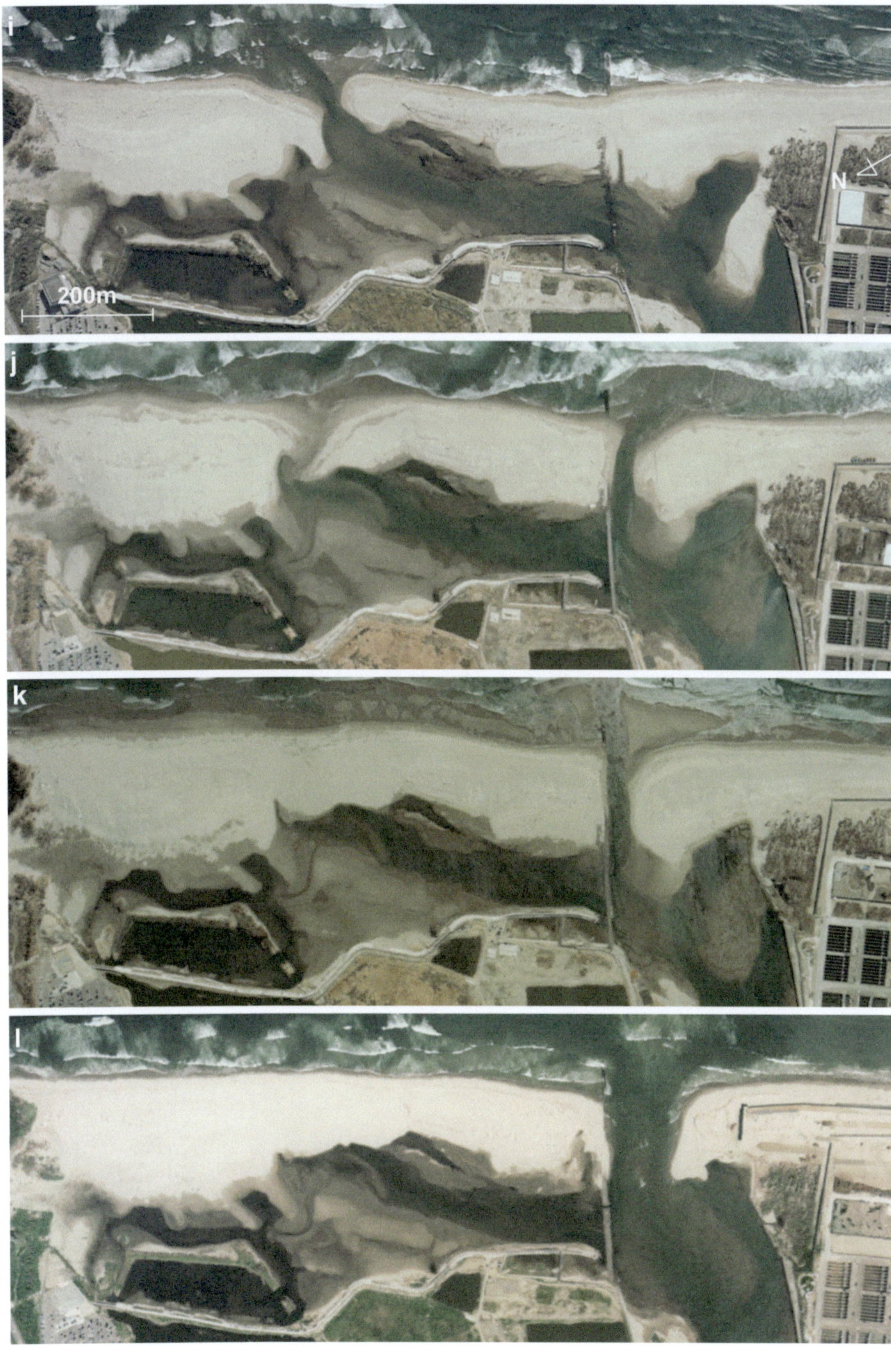

Fig. 11.2 (continued)

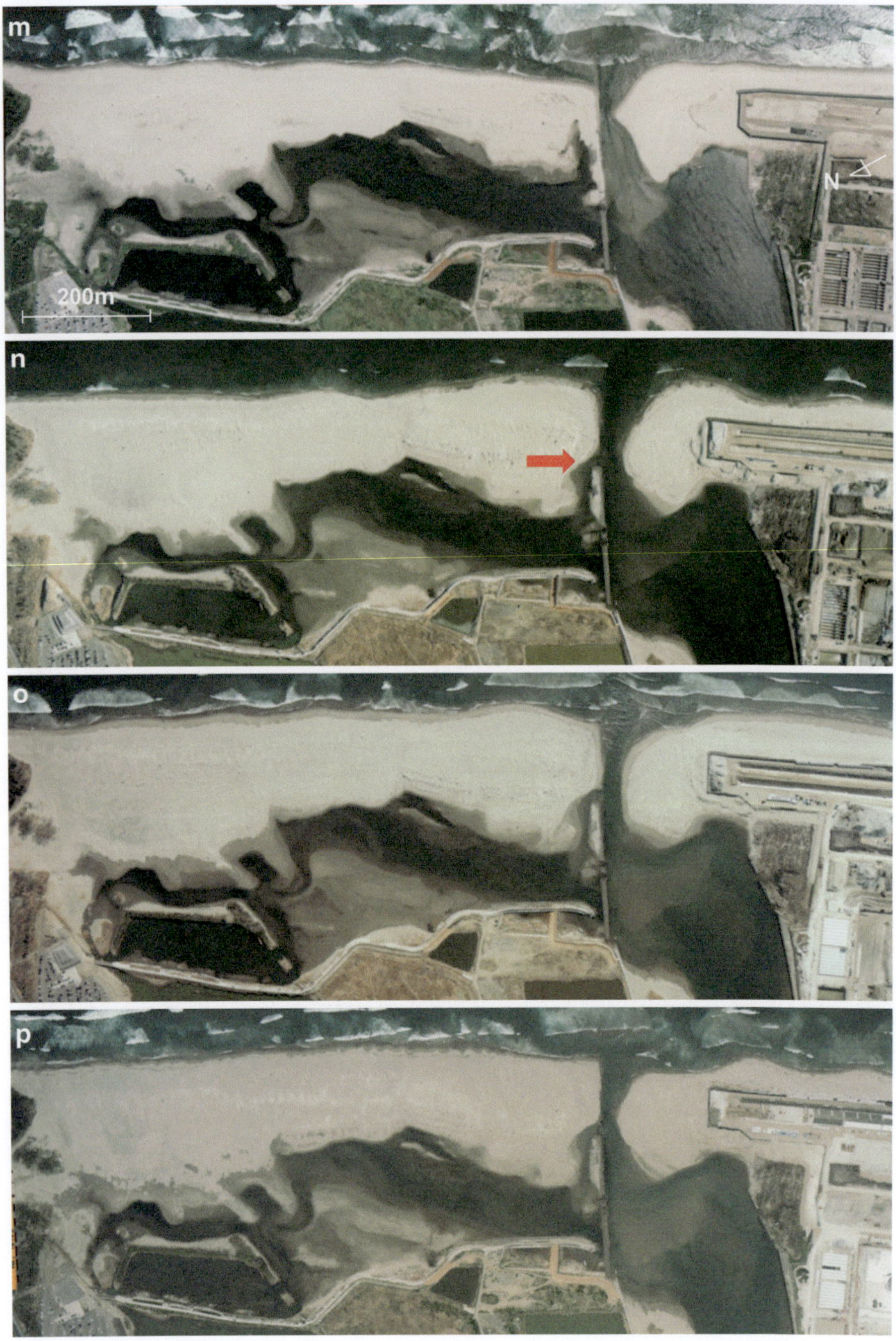

Fig. 11.2 (continued)

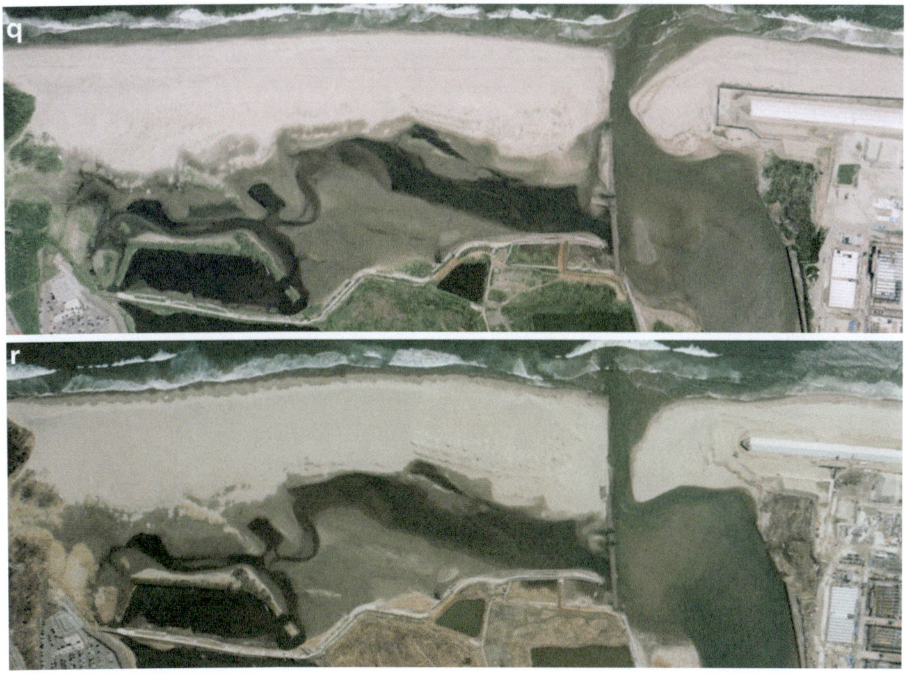

Fig. 11.2 (continued)

11.3.2 Morphological Changes Induced by the Great East Japan Earthquake and Tsunami

Figure 11.2b shows the morphology of river mouth area immediately after the tsunami hit. The photograph was taken on March 12, 2011, one day after the event. The sandy barrier in front of the lagoon was completely eroded, only a few small islands from the body of this barrier still remained. Moreover, the deposition of the sand in the former water area of the lagoon was observed. The above mentioned jetty can be seen clearly on the left hand side of the river mouth, and in particular, the severe retreat of the shoreline on the left side of the river mouth was remarkable.

11.3.3 Morphological Recovery After the Great East Japan Earthquake and Tsunami

Aerial photographs in Fig. 11.2c–e were collected from Google Earth. These figures show the recovery of morphology in the first 2 months after the tsunami. Through this period regular aerial photograph taking was unfortunately still suspended. The aerial photograph in Fig. 11.2c, taken about 2 weeks after the tsunami, shows that the morphology of the lagoon was not so different from the morphology right after the tsunami (Fig. 11.2b). However, the erosion of sandy beach on the right side of the Nanakita River mouth can be observed. The photograph in Fig. 11.2d indicates that the sandy beach divided by the tsunami to a few islands was almost connected about 3 weeks after the tsunami. However, the water area in the lagoon was still occupied by large amount of deposited sediment. Furthermore, more severe erosion of sandy beach on the right side of the Nanakita River mouth can be observed.

Figure 11.2e shows that shoreline in the lagoon area has been connected. The erosion of sandy beach on the right side of the Nanakita River mouth was still happening and getting more serious. Aerial photographs shown in Fig. 11.2f, g, illustrate clearly the development of new sand spit at the river mouth. However, its location was shifted upstream. The location of a tip of the new sand spit was located about 400 m upstream compared to that before the tsunami (Fig. 11.2a). Similar phenomenon has been also observed at other affected river mouths. The tsunami scoured the river mouths becoming deeper than the depth of closure, and hence waves could also propagate upstream and caused the intrusion of sand spits (Tanaka et al. 2013). Subsequently, the closure of Nanakita River mouth was confirmed on August 10, 2011. The aerial photograph taken after the closure is presented in Fig. 11.2h. It is important to note that the closure of this river mouth has been occurred frequently through the 1990s (Tanaka et al. 1996). After the river mouth closed the flow was diverted to Natori River, which is located about 8 km south of Nanakita River, through the Teizan Canal. Due to this reason the flushing of the sandy barrier to reopen the river mouth did not occur. Moreover, the river mouth morphology remained stable as seen in Fig. 11.2h.

The flood induced by the typhoon No. 15 on September 22, 2011 caused the breaching of the sandy barrier by the corresponding place where the former river mouth was located. Hence, a new river mouth was formed as it can be observed in Fig. 11.2i. The curved river channel, which flows inside and through the former water area of Gamo Lagoon, raised the water level during the flooding and altered the brackish water environmental conditions too.

After the strengthening of the jetty separating Gamo Lagoon and the Nanakita River mouth in early March, 2012, an artificial excavation was carried out at the location of the former Nanakita River mouth (Fig. 11.2j). The excavated sand was then added to the sandy barrier on the left side of the mouth. Once the excavation finalized the morphology returned to the previous condition similar to that before the closure, and the position of the river mouth was stabilized at the same place before the tsunami hit (Fig. 11.2k).

Figure 11.2l was taken in the summer of 2012 showing that the width of the Nanakita River mouth was increasing in turn due to the increase of the river discharge. Subsequently, when the discharge decreased through autumn and winter its mouth width decreased extremely (Fig. 11.2m–o). According to Fig. 11.2n–q, a part of jetty on the left side of the river mouth (pointed by a red arrow) was destroyed allowing the sea water to directly flow into the lagoon raising concerns about the increase in the salt concentration, or changes in the ecosystem. Subsequently, this portion was artificially blocked as of February 24, 2014 (Fig. 11.2r). Moreover, the pattern of the river mouth sand spit does not appear to be different when compared with that before the tsunami stroke.

A big change in the water area of Gamo Lagoon does not seem to have occurred since the photograph was taken in June, 2011 (Fig. 11.2f). In order to confirm this situation, the shoreline and waterline positions have been extracted from the photographs presented in Fig. 11.2 and the variation of both are presented in Fig. 11.3. According to the results shown in the latter figure, a gradual recovery of the shoreline could be observed from the severe retreat induced by the tsunami.

On the other hand, the large change of the waterline position has not been observed since June 8, 2011. The temporal variation of the shoreline and waterline positions on two transects seen in *L*1 and *L*2 (Fig. 11.3) are also illustrated in Fig. 11.4. As previously mentioned all the shoreline positions are corrected to the calculated tidal level at Sendai Port and the averaged beach slope. However, for the waterline, it is difficult to make such correction. Hence, waterline data has been directly utilized after extracting them from the aerial photographs. The analysis has shown that the stable trend of waterline and shoreline positions in the recovery process can be confirmed as mentioned above.

The seasonal variation of waves also has an impact on the variation of shoreline position. When the shoreline is reaching a stable stage, the impact of the seasonal variation of waves on the variation of shoreline position can be observed as from the 400th day after the tsunami took place.

Figure 11.5 shows the water area (water surface inside the lagoon), $A_{W,}$ of Gamo Lagoon when extracted from the aerial photographs shown in Fig. 11.2. The analysis showed that although the tidal level variation has an influence on the lagoons area, the net reduction of about 40 % when compared with that before the tsunami took place can be observed.

11.4 Discussion

Before this study, Liew et al. (2010) utilized 1 m-resolution yearly satellite images (IKONOS) to investigate the recovery of morphology of the coasts in Indonesia and Thailand after the 2004 Indian Tsunami. That study reported a case of severe damages of barrier beach with lagoon. The lagoon was opened directly to the sea after the tsunami. The recovery of morphology in that area was rather rapid, about 1 year after the tsunami the new beach has been built. This indicates that the results carried out from present study show similar trend of morphology changes and recovery with the one in that study. In addition, Liew et al. (2010) also pointed out that although the width of new beaches tends to be larger than previous, shoreline positions have not reached the one before the tsunami. It is consistent with the recovery situation of the beach in front of Gamo Lagoon area. Hoang et al. (in preparation) investigate the recovery of concave shoreline at this study area after the tsunami. According to that study, rigid boundaries (structures, breakwater of Sendai Port on the north side and drainage on the south side of lagoon) on both sides of concave

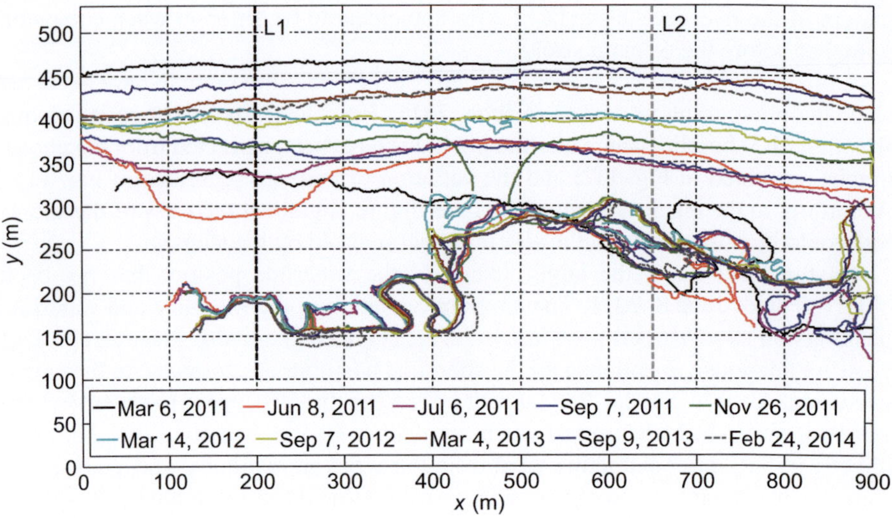

Fig. 11.3 Changes of shoreline and waterline positions at Gamo Lagoon (Origin of coordinates in this figure is represented in Fig. 11.2a)

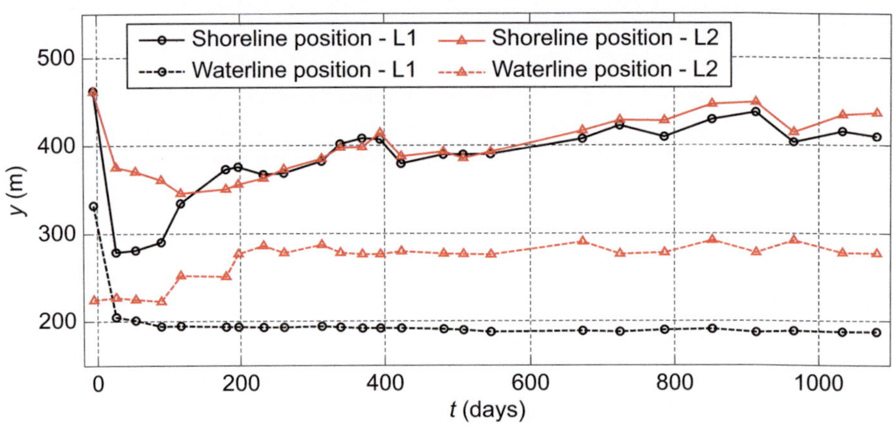

Fig. 11.4 Temporal variation of shoreline and waterline positions at Gamo Lagoon

portion make shoreline getting new equilibrium earlier than the case without rigid boundaries. However, the equilibrium position is more landward compared to its position before the tsunami. In the case of beaches reported in Liew et al. (2010), they have no coastal structures while there are rocky headlands on both ends. These headlands are considered as rigid boundaries.

In the present study, the detailed analysis on the changes and recovery of morphology is made while that could not be done in previous studies. That analysis could be carried out based on the high resolution, and especially very frequent photograph taking. This is one of the most advantages of the approach used in this

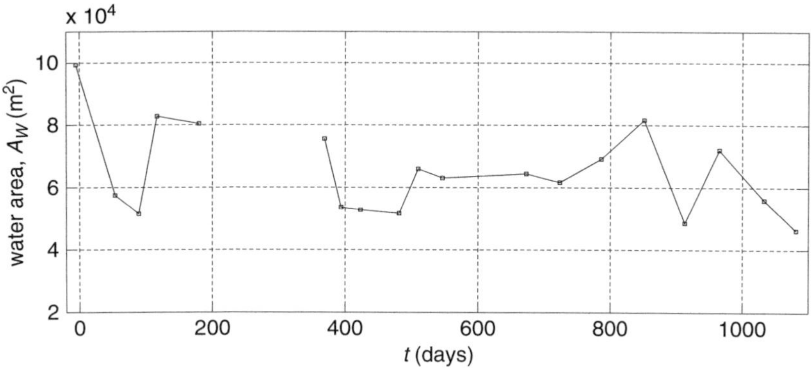

Fig. 11.5 Temporal variation of water area of Gamo Lagoon

study. However, the high resolution and frequent taking aerial photograph is not always available for every area. Hence, these matters could be the difficulty encountered when applying this approach to other areas.

11.5 Conclusions

The morphology of Gamo Lagoon was severely damaged by the 2011 Great East Japan Earthquake and Tsunami. Aerial photographs were utilized to analyze the topographical changes induced by the tsunami and subsequent recovery process. Results showed that there is a significant reduction in the lagoon water area, and also a change in the shoreline process. The recovery of the ecosystem in coastal lagoons, like Gamo Lagoon, is linked to the changes and the recovery of the lagoon's morphology, and therefore to further assess these changes and also its recovery, further monitoring and detailed studies are required to continue in future.

Acknowledgement This work was supported by River Environmental Fund (REF) from the Foundation of River and Watershed Environmental Management (FOREM). Authors would like to express their gratitude to the above support.

References

Ali PY, Narayana AC (2015) Short-term morphological and shoreline changes at Trinkat Island, Andaman and Nicobar, India, after the 2004 tsunami. Mar Geod 38(1):26–39

Choowong M, Phantuwongraj S, Charoentitirat T, Chutakositkanon V, Yumuang S, Charusiri P (2009) Beach recovery after 2004 Indian Ocean tsunami from Phang-nga, Thailand. J Geomorphol 104(3–4):134–142

Elizabeth HB, Ian LT (2005) Shoreline definition and detection: a review. J Coast Res 21(4):688–703

Goto K, Hashimoto K, Sugawara D, Yanagisawa H, Abe T (2014) Spatial thickness variability of the 2011 Tohoku-oki tsunami deposits along the coastline of Sendai Bay. Mar Geol 358:38–48
Hoang VC, Mitobe Y, Tanaka H (2014) Analysis of shoreline behavior on Sendai Coast before and after the 2011, Proceedings of 34th international conference on coastal engineering
Hoang VC, Tanaka H, Mitobe Y (2015) Theoretical study on the recovery process of the concave landform after the tsunami. J Jpn Soc Civil Eng Ser B1 (Hydraul Eng) 71(4)
Hoang VC, Tanaka H, Mitobe Y (in preparation) Analytical solution on restoration of eroded beach morphology after the 2011 tsunami
Liew SC, Gupta A, Wong PP, Kwoh LK (2010) Recovery from a large tsunami mapped over time: the Aceh coast, Sumatra. J Geomorphology 114(4):520–529
Malloy KD, Yamashita Y, Yamada H, Timothy E (1996) Spatial and temporal patterns of juvenile stone flounder *Kareius bicoloratus* growth rates during and after settlement. Mar Ecol Prog Ser 131:49–59
Minoura K, Imamura F, Sugawara D, Kono Y, Iwashita T (2001) The 869 Jogan tsunami deposit and recurrence interval of large-scale tsunami on the Pacific coast of northeast Japan. J Nat Disaster Sci 23(2):83–88
Moore LJ (2000) Shoreline mapping techniques. J Coast Res 16(1):111–124
Omori M, Kinno H, Nishihata I (1976) Study of the habitat of juvenile stone flounder, *Kareius bicoloratus* (BASILEWSKY), in the Estuary of the Nanakita River. Tohoku J Agric Res 27(2):79–91
Pradjoko E, Tanaka H (2010) Aerial photograph of Sendai Coast for shoreline behavior analysis. Proceedings of the 32nd international conference on coastal engineering
Sakamaki T, Nishimura O, Sudo R (2006) Tidal time-scale variation in nutrient flux across the sediment–water interface of an estuarine tidal flat. Estuar Coast Shelf Sci 67(4):653–663
Shin SW, Nishimura O (2013) Utilization of food sources before and after the tsunami in Nuttallia olivacea at Gamo Lagoon, Japan. Environ Eng Res 18(4):259–265
Takashimizu Y, Urabe A, Suzuki K, Sato Y (2012) Deposition by the 2011 Tohoku-oki tsunami on coastal lowland controlled by beach ridges near Sendai, Japan. Sediment Geol 282:124–141
Tanaka H, Takahashi F, Takahashi A (1996) Complete closure of the Nanakita River mouth in 1994. Proceedings of the 25th international conference on coastal engineering, pp 4545–4556
Tanaka H, Suntoyo, Nagasawa T (2002) Sediment intrusion into Gamo Lagoon by wave overtopping. Proceedings of the 28th international conference on coastal engineering
Tanaka H, Tinh NX, Umeda M, Hirao R, Pradjoko E, Mano A, Udo K (2012) Coastal and estuarine morphology changes induced by the 2011 Great East Japan earthquake tsunami. Coast Eng J 54(1):1250010 (25 pages)
Tanaka H, Bagus MA, Roh M, Hoang VC (2013) Sand spit intrusion into a river mouth after the Great East Japan tsunami. J Jpn Soc Civil Eng Ser B2 (Coast Eng) 69:I_616–I_620 (in Japanese)
Tanaka H, Adityawan MB, Mano A (2014) Morphological changes at the Nanakita River mouth after the Great East Japan Tsunami of 2011. Coast Eng 86:14–26
Tappin DR, Evans HM, Jordan CJ, Richmond B, Sugawara D, Goto K (2012) Coastal changes in the Sendai area from the impact of the 2011 Tōhoku-oki tsunami: Interpretations of time series satellite images, helicopter-borne video footage and field observations. Sediment Geol 282(30):151–174
Tinh NX, Tanaka H (2011) Discussion of overwash prevention construction on the northern part of Sendai Coast. Proceedings of 6th international conference on coastal structures. (CD-ROM)
Udo K, Sugawara D, Tanaka H, Imai K, Mano A (2012) Impact of the 2011 Tohoku Earthquake and Tsunami on beach morphology along the northern Sendai Coast. Coast Eng J 54(01):1250009 (15 pages)
Yamashita Y, Otake T, Yamada H (2000) Relative contributions from exposed inshore and estuarine nursery grounds to the recruitment of stone flounder, Platichthys bicoloratus, estimated using otolith Sr:Ca ratios. Fish Oceanogr 9(4):316–327

Chapter 12
The Minato River in Miyagi Prefecture Reconstruction and Restoration – An Overview

Vicente Santiago-Fandiño and Naoko Kimura

Abstract On March 11, 2011 (Heisei 23) a large earthquake and tsunami hit the coastline along Miyagi Prefecture, Japan, with catastrophic consequences by heavily impacting beaches, lagoons, wetlands, rivers and estuaries. The lower watershed of the Minato River (Minatogawa) and its mildly urbanized estuary, as well as the surrounding area, suffered heavy damage due to ground subsidence, sediment deposition, wave erosion, abrasion and flooding. Local authorities developed plans for reconstruction and rehabilitation of the river and its estuary. Work started in earnest by 2013, which included the construction of a L1 type levee. The impacts of the engineering-related works and structures to be built are expected to be large, particularly for the levee, likely putting in doubt their environmental sustainability.

Keywords Minato River Minatogawa • Restoration reconstruction Miyagi

12.1 Introduction

On March 2011 a 9.0 M_w earthquake created a megatsunami that hit the Sanriku Coastline along Iwate and Miyagi Prefectures. Tsunami waves, their run-up and backwash, generated turbulence and currents leading to large erosion and abrasion resulting in massive destruction, environmental and soil alteration as well as sediment deposition (Shi 2003) capable of reshaping the coastal landscape (Dawson 1994). Estuaries, lowland rivers, as well as unprotected dry riverbeds allowed

---

V. Santiago-Fandiño (✉)
Environmental Advisor, El Curbiellu 28, Villaviciosa, Asturias 33314, Spain
e-mail: v.santiago.f@gmail.com

N. Kimura
Department of Sustainable Rural Development, Graduate School of Global Environmental Studies, Kyoto University, Kyoto, Japan
e-mail: kimura.naoko.34u@st.kyoto-u.ac.jp

V. Santiago-Fandiño et al. (eds.), *Tsunamis and Earthquakes in Coastal Environments*, Coastal Research Library 14, DOI 10.1007/978-3-319-28528-3_12

tsunami waves to flow inland causing flooding (Shimazu 2012). The tsunami also displaced or broke coastal and river-protecting structures such as concrete tetrapods, seawalls, levees and other structures. Moreover, ground subsidence and displacement from the earthquakes added to the damage (Kamiyama et al. 2012; Japan Water Forum 2012).

Major changes in water surface area, course, flow, depth, sediment quality, and bottom sediments in rivers and lagoons were common. Flora and fauna were likewise heavily impacted resulting in biodiversity changes and ecotone destruction (Biodiversity Center of Japan 2013). Shadanhojin Tohoku Kensetsu Kyokai (2012) published an extensive photographic record of the coastal morphological changes along Iwate, Miyagi and Fukushima Prefectures.

Large numbers of studies were undertaken along the Sanriku coastline after the March 2011 events; Santiago-Fandiño (2013) and Santiago-Fandiño and Kim MH (2015) gave an overview of the vast destruction of settlements, agricultural fields, forestry, fisheries and infrastructure, as well as the resulting potential contamination from tsunami debris in Miyagi. Komatsu et al. (2013) analyzed the erosion undertaken by tsunami waves on natural landscapes in the city of Aneyoshi in Iwate prefecture while Udo et al. (2015) did the same for the southern Sendai coast of Miyagi. Shimazu (2012) studied the geomorphic impacts in the lower reaches of the Natori River and Grzelak et al. (2013) determined the ecological status of sandy beaches after the 2011 tsunami focusing on meiofauna in Sendai, too.

Furthermore, Roh et al. (2014) studied the ascending wave propagation of the March 2011 tsunami on the Kitakami River course; Nagasawa and Tanaka (2012) described the structural damage and geomorphic changes caused at various places in Iwate and Miyagi prefectures, while Tanaka et al. (2013, 2014) examined the relationship between river morphology and tsunami propagation in rivers in Miyagi. Adityawan et al. (2012) studied the degree of tsunami wave intrusion on river mouth morphological features, such as bed slope, in the same prefecture.

Rivers in Japan are classified into three main groups, i.e., *Class A* (109 rivers), which include 13,798 rivers systems with a total length of 87,172 km; *Class B* (2691 rivers), including 6931 river systems (35,717 km); and *Locally Designated rivers*. Atsumi (2009) describes, however, 2726 Class B rivers. Also, an alternative river classification system based on landscape and ecological aspects resulted in six groups instead of three (Nakagoshi and Inoue 2003). Class A rivers are administered by the Ministry of Land, Infrastructure, Transport and Tourism, Japan (MLIT), and Class B by prefectures and municipalities (Omachi 1999 and MLIT 2014a).

Miyagi Prefecture contains four large Class A rivers, i.e., Kitakamigawa (Fig. 12.1), Narusegawa, Natorigawa and Abukumagawa (MLIT 2014a); 68 Class B rivers (as of March 2001), including the Minato River (MPG 2011; Wikipedia-Japan 2014), which were all affected by the tsunami and earthquake events in March 2011. The Tohoku Regional Bureau (2015) has reported that a number of estuarine levees suffered tremendous damage as those on the Kitakamigawa and the Abukumagawa rivers. Moreover, this agency also noted that most of those levees constructed south of Sendai Bay were badly deformed or washed away for almost 50 km along the

Fig. 12.1 Kitakamigawa River uppershed in Tome City (*left*) and the estuary (*right*)

coastline. Oka et al. (2012) concluded that as result of the 2011 event about 146 rivers suffered embankment damage at 611 sites in Miyagi prefecture, while additional structural damage occurred at 25 sites on 21 rivers. The Minato River and estuary was not an exception in the amount of damage sustained.

12.2 Minato River

The Minato River in the north of Miyagi prefecture is located about 470 km north of Tokyo (Fig. 12.2). A typical Ria,[1] this river flows from the Nana-une Mountains (MT 2015) to the Pacific Ocean through a nearby estuary where Utatsu-Namiita village is located (38° 44′ 33″ N- 141°32′ 27″) in Minamisanriku Town. The mainstem is 2.5 km in length but if the tributaries are included then it reaches 6 km in total (MT 2015); its basin covers 3.36 km^2. Minato Port that is dedicated to fisheries and aquaculture, is shared by Utatsu-Minato and Utatsu-Namita villages (Fig. 12.3). Forested hills with a variety of pine trees and occasional rice fields are found along its banks.

A small dam to control sediment and energy flow in Minato River was built 52 m above sea level and 1.8 km from its source in the Nana-une Mountains. This structure also serves as the divide between the upper and lower watersheds (Figs. 12.4 and 12.6).

For many years as result of the implementation of a use and protection rivers policy (Godou 2010), the river and estuary have gone through urbanization, land reclamation, embankment construction, revetment installations and canalization (Fig. 12.3) all impinging on natural sedimentation and land-water interactions, including nutrient exchange and natural purification in the ecotone and riparian environments. Clilverd (2015) has shown that embankments and river canalization produce extensive ecological degradation in rivers throughout the world.

[1] A partially submerged river valley mostly surrounded by mountains with a long, narrow inlet.

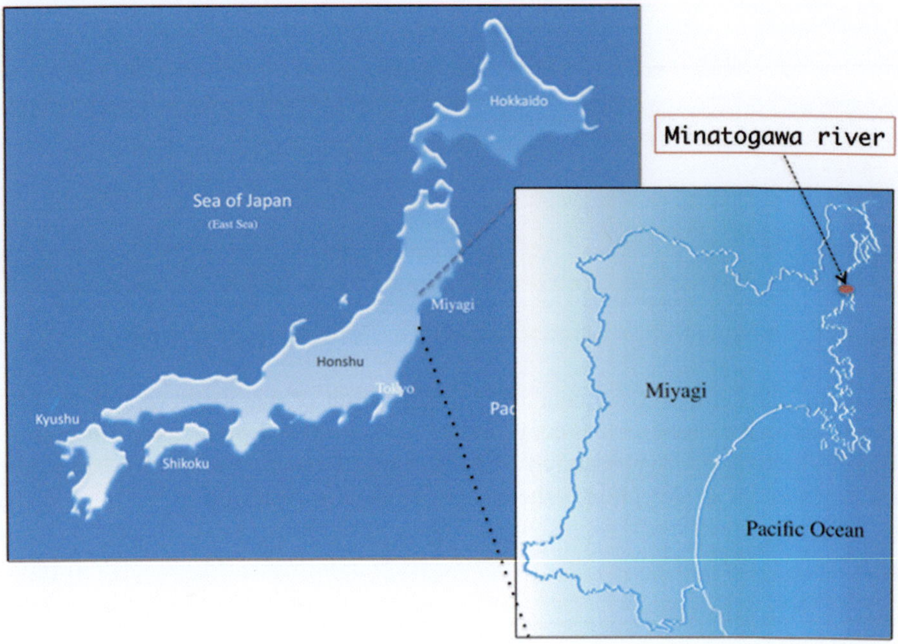

Fig. 12.2 Minatogawa River location; north east coast of Miyagi Prefecture. Japan

12.2.1 Tsunami and Earthquake Environmental Impacts

The Sanriku coastline has been hit by a number of large earthquakes and tsunamis throughout its history (Shuto and Fujima 2009) that have impacted the rivers, estuaries and coastal ecosystems. The most recent damaging ones were the Showa (March 3, 1933) and the Great East Japan Earthquake and Tsunami, or Heisei Earthquake Tsunami[2] (Imamura and Anawat 2012) or GEJET (Japan Red Cross Society 2013) of 2011, both having had run-ups reaching 28.7 m and 38.9 m, respectively (Mori et al. 2011).

Miyagi Prefecture was hit by a series of tsunami waves in 2011, the second appears to have been the largest in Utatsu Minato and Utatsu-Namiita with run-ups between 19.4 and 19.5 m and 14 and 14.5 m respectively (Tsuji et al. 2014). The wave flowed deep into the Minato river estuary (Figs. 12.5 and 12.6).

Figure 12.6 shows the approximate inundation area of the March 2011 tsunami along the Minato lower watershed, covering about 235,000 square meters within a

[2] Although the Chilean Tsunami of 1960 also had a large impact it can be considered as minor when compared to the Showa and Heisei events.

Fig. 12.3 Minatogawa Bay and Port as seen in 2009. The largely modified river and estuary, rice fields and forested hills can be easily identified. Source: Google Map

Fig. 12.4 Early Minatogawa River uppershed sediment control dam

Fig. 12.5 First tsunami wave flowing through the Minatogawa River reaching Route 45 (photos provided by a local citizen), and the 3 m fully submerged underpass

perimeter of about 65,000 m as calculated from the map.[3] Local inhabitants pinpointed the highest run-up site at about 1 km away from the river mouth calling it the "Harai no Chi" mark (38°44′19.05″N, 141°31′11.06″E). A difference of about 200 m inland from this point is observed when compared to that of the tsunami inundation map produced by the Association of Japanese Geographers,[4] likely composed from aerial photos and satellite images alone. A memorial stone of the Showa tsunami describing its impacts and casualties was relocated inland from Minato Port to nearby Route 45 for conservation purposes.

A satellite image (Fig. 12.7) shows the widespread damage sustained by the Minato River due to the 2011 tsunami. Most of the lowland settlements, port facilities, sea fences, embankments and revetments, as well as rice fields, were badly damaged or destroyed. Figure 12.8 shows this damage; direct wave impact, currents, and turbulence, as well as backwash, were the most destructive forces. Moreover, ground subsidence, tsunami debris and mud accumulation further added to the environmental damage altering soil quality, salinity and porosity, texture and permeability, which in turn may have affected groundwater flow.

---

[3] The inundation map was produced in ArcGIS (ESRI). Flooded areas were referred to those in the tsunami inundation map by the Association of Japanese Geographers-AJG- (http://map311.ecom-plat.jp/map/map/?mid=40). The basic geospatial information (*kibanchizu-joho*) was taken from the Geospatial Information Authority of Japan (GSI/MLIT) (http://fgd.gsi.go.jp/download/).

[4] http://map311.ecom-plat.jp/map/map/?mid=40

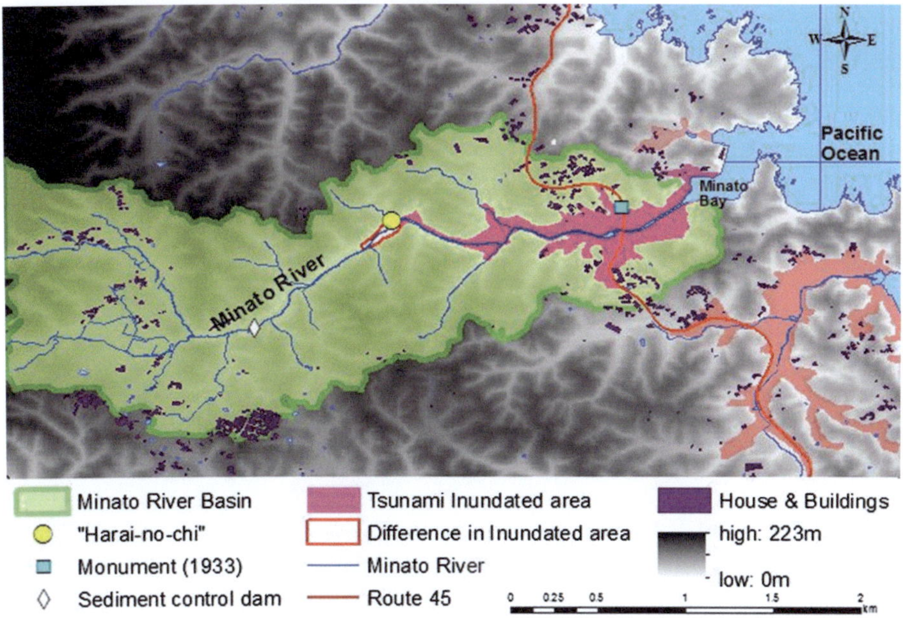

Fig. 12.6 Minatogawa River Basin and tsunami inundation map; the *yellow dot* represents the "Harai-no-Chi" mark

The river's surface flow, depth and course were altered as stones, gravel, rubble and concrete slabs were wide spread. Moreover, uprooted vegetation from nearby hills, e.g., black and red pine (*P. thunbergii and P. densiflora*) and Japanese cedar (*C. japonica*) trees and other vegetation accumulated in some areas (Fig. 12.8). The fact that ground subsidence and displacement took place further aggravated the impacts.

On the hillsides, ground and soil instability became common, not only as result of the tsunami impacts and its related forces, but because the frequent tremors, seasonal rains and typhoons generated landslides that added to the environmental degradation (Fig. 12.9).

12.3 Environmental Recovery

The speed of natural recovery in heavily impacted tsunami areas may vary depending on the severity of the damage, the particular type of ecosystem(s), loss of habitats and affected species. Although it is difficult to make a meaningful assessment of the natural restoration process of the river due to time availability and lack of dedicated studies, by visiting the area once a year from April 2011 to early 2015 a general idea of the ongoing changes was obtained.

Fig. 12.7 The Minatogawa River and estuary after the 2011 tsunami showing the scale of devastation along the lowershed (Source: Google Earth)

The sea appeared to have reclaimed some of the estuary's natural flooding areas, as well as riverine sections from urbanization and agriculture. Moreover, most of the original pre-tsunami existing habitats were destroyed or badly damaged, which in turn provided new grounds and environmental changes, such as water and sediment quality, whereby pioneering or opportunistic species started to grow profusely, e.g., from incipient riparian vegetation and algal growth along the river (Fig. 12.10) to a more widespread growth almost covering former rice fields (Fig. 12.11). The dilution and wash out of sea salt from soil and ponds due to seasonal rains, snowmelt and typhoons accelerated the transformation process as it also facilitated soil reconditioning and microfloral growth, paving the way for the appearance of primary ecosystems.

Other opportunistic species, such as seagulls, colonized eroded bare ground, serving also as food sources for predators such as foxes and birds of prey (Fig. 12.12). Some aquatic species started to reappear, such as fish and crustaceans, since they were also abundant in the river. Reports by local fisherman include Ayu fish (*P. altivelis*), Japanese dace –Ugui- (*T. hakonensis*), Ice fish (*S. microdon*), chum salmon –Aki-zake - (*O. keta*) and others, all included in the ichthyofauna of Tohoku published by Matsuura et al. (2009). Although similar to chum salmon, silver salmon (*O. kisutsh*) is the only species used for aquacultural purposes; a facility was constructed by a local fisherman on the river side as part of the restoration process (Fig. 12.13).

Fig. 12.8 Minatogawa River degradation along the lowershed and estuary a month after the tsunami stroke

Fig. 12.9 Landslides and ground sinking along the Minatogawa River lowershed

Fig. 12.10 Riparian and aquatic vegetation growth along the river and estuary after March 11 events

12.4 Reconstruction and Restoration

Soon after the earthquake and tsunami an emergency response plan was put into action to provide relief to isolated houses and villages. As part of this process, debris cleanup and collection started along the damaged watershed (Fig. 12.14) and became the precursor of the restoration process. At the same time, sections of the riverbank started to be reinforced with sand and/or debris bags in order to reduce erosion (Fig. 12.17).

The Miyagi Prefectural Government, in collaboration with the local authorities and villagers, developed a detailed restoration (recovery) and reconstruction plan in 2012. At the time, the Japan Railways (JR) Company already had considered the construction of a bus route using the former JR Sanriku railway line destroyed by the 2011 tsunami; the service started tests in 2014 (Fig. 12.15).

The overall government plan considered, amongst other measures, sea fences, settlement relocation to higher ground, creation of local roads and river restoration, which in turn included the construction of a levee (designed to provide flood protection by containing, diverting and controlling water (NAS 2012).

Levee construction started in late 2014 (Fig. 12.16) and expected to be finished by the end of 2015 or early 2016. The structure incorporates two embankments and

Fig. 12.11 Pioneering and or invasive vegetation in former rice fields

Fig. 12.12 Hawk feeding on a recently killed seagull from a colony settled by the estuary

Fig. 12.13 Silver salmon fry aquaculture facility built on the Minatogawa riverside. A large accumulation of tsunami mud and debris on the exposed riverbed can be spotted

Fig. 12.14 Early debris cleanup works along the Minatogawa River and effluents

two sluice gates to control the water flow. One embankment will be 930 m long with a smaller one of only 660 m.

Both embankments will have a height of 9.80 m for the first 412 m from the river mouth and be reduced to 5.6 m thereafter to the end. The width of both structures will vary between 15 and 55 m (MPG 2014). Moreover, to facilitate the river's discharge its course will be straightened in some sections by the levee's design.

Work such as drilling, soil compacting and ground conditioning as well as the creation of soil mounds for the construction of the embankments continued through

Fig. 12.15 Remains of the former JR Sanriku railways line and the new JR bus line crossing over the Minatogawa River

2015. Likewise, reallocation work to build new housing in the nearby hills, forest clearance and the construction of a few local new roads is taking place (Fig. 12.17).

Building and construction works create serious environmental problems; they increase soil erosion and run-off, damage or destroy the ecotone as well as damage riparian vegetation, and heavily affect aquatic fauna while increasing habitat loss and degradation. Digging river beds to allow for faster water discharge heavily affects the microfauna and concrete embankments stop nutrient exchange between land and water that are natural water purification processes that affects water quality. Last but not least, ground water recharge is altered and erosion and sedimentation increase as water velocity increases (SAFOA and US Army Corps of Engineers 2015; Inoue and Kamogawa 2012).

12.5 Discussion and Conclusions

River management policies in Japan have gone through a series of revisions from 1896, when the first River Law was enacted focusing on flood prevention and control, to incorporate water use in 1964. In 1997 an amendment was made to the River

Fig. 12.16 Levee construction as of March 2015. Wide roads, concrete slabs, construction and drilling equipment as well as soil mounds are common sightings along the Minatogawa River shores

Law to include conservation and improvement of the river environment. Moreover, in 1999 the Sea Coast Law was amended to provide protection for seacoast environments (Godou 2010; Omachi 1999; and Nakamura et al. 2006). Yoshimura et al. (2005) concluded that despite these efforts rivers were still heavily impacted due to the loss of dynamic flood plains, canalization, flow regulation and intensive urbanization. Nonetheless, Hirano (2013) noted that by 2015 more than 23,000 river restoration projects were undertaken over the last 15 years as part of the "Nature-oriented River Works" initiative of 1991–2001.

In the case of the Minato River and estuary the transformation resulting from the implementation of river management policies is conspicuous, with riverine and estuarine conditions being drastically modified to control sediment flow from the upper watershed by construction of a small dam. Likewise, many coastal and riverine sections went through heavy urbanization and land use change that included the development of rice fields.

Fig. 12.17 Environmental degradation along the river caused by levee construction and housing relocation including soil compacting, forest clearance and shores alteration

Although the Kathleen and Ione typhoons in 1947 and 1948, respectively, and the Chilean Tsunami of 1960 certainly had an impact in Miyagi (Nakamura et al. 2008, Ellis and Sherman 2015, Nakamura 2015 and JCOLD 2009) and hence on the Minato River and estuary too, the recent relative environmental stability was seriously affected by the 2011 Heisei event. The magnitude of the reconstruction and restoration work and the levee construction following this recent event will have lasting environmental impacts on the Minato River. The levee, measuring 5–10 m high, almost one kilometer long and 15–55 m wide, will certainly become the most important impacting factor on the estuary and overall on the river's lower watershed (S&TT 2012).

Aside from the visual effects and environmental issues, which this structure poses to the environment and landscape, some inhabitants complain about the capacity of this levee to offer full protection against flooding. Coastal levees in Japan are now designed and classified considering tsunami frequencies and heights. L1 protect against events reoccurring between 100 years or longer with heights between 5 and 15.5 m, and L2 are designed for those that may occur between 500 and 1000 years and measure >15.5 m, as was the case with the Heisei tsunami (Iuchi et al. 2014; Hirano 2013). Considering the fact that the March 2011 tsunami exceeded 16 m, reaching up to 19.4–18.5 m at Utatsu-Minato and Utatsu-Namiita villages, respectively, on the Minato river estuary (Tsuji et al. 2014), the selected L1 levee design does not seem adequate to protect against this type of large and distant event but instead requires a L2 design. Nevertheless, and although it is impossible to predict with certainty the levee's efficiency, the L1 design option seems to be adequate to protect against most frequently occurring tsunamis of 5–15.5 m height as well large sea surges and high tides impacting this area.

Some of the controversial points related to the building of this levee by the authorities will be related to its high maintenance costs as these types of structures tend to decay with time, as well as the opposition by some local residents on its construction, size and purpose.

Parallel impacts related to the restoration and reconstruction of the Minato River occur due to the need to bring construction materials such as soil from other sites creating a variety of environmental problems somewhere else. The same could be said about the transportation of materials to build cement and concrete slabs, revetments and other structures. Although Taisei Corporation received a 2013 award for Environmental Winning Technology for its development of an aggregated cement-hardened material using concrete rubble created by the March 2011 events. At the time this chapter was written it was not clear that this technology was being used in the construction of the Minato River levee as the plans only mention the use of soil as filling material (earth embankments), even though the new material was promoted as a concrete substitute for levee filling material (Taisei Corporation, 2013).

In Japan the coastal policy calls for the preservation of habitats, care for the scenery and use of beaches coupled with the protection from high tides and sea surges as well as tsunamis, all in balance with factors such as defense, environment and usage (MILT 2014b). Oddly, it appears that the current decision to restore and rehabilitate the Minato River and estuary is basically directed towards protection and usage with scant consideration about its real environmental sustainability.

Ironically, the Heisei tsunami and earthquake has given a unique opportunity for decision makers to come up with more sustainable environmental options as many scientific reports are concluding that the fauna and flora are strongly coming back to the tsunami and earthquake affected sites. In some places there is the potential to have even better productivity than before the event occurred. But instead of following and supporting this process a classical engineering approach continues to be employed that severely damages the environment and which has been shown to impact many coastal lentic and lotic systems, including the Minato River. Although it is certain that certain types of environmental beautification will be made along the embankments to make them more appealing to the local population and enhance the landscape, these will produce a sustainable, ecologically-aesthetic restoration greatly unlike the natural environment in the newly restored Minato River.

There are a large number of altered ecosystems in the Anthropocene era, which despite all kinds of restoration efforts, will never return to their original state. Moreover, it may no longer be appropriate. There is a need to look at what is possible to change while also accepting what is not possible. Unless humans change the way they manage natural resources and landscapes new types of ecosystems will have to be accepted as well as their human-modified baselines. In managing nature the points of reference must take into account human effects while ensuring that these do not cause further environmental degradation Keller et al. (2015).

Acknowledgements The present study was made possible due to the kind support of Mr. M. Nakayama, President and CEO of Nakayama Industries in Shiga Prefecture and Mr. K. Sakakibara from Tome City, Miyagi Prefecture in Japan. We also wish to thank the reviewer for valuable comments and suggestions. Last, but not least to R.D. Robarts for proofreading the text.

Photos taken by V. Santiago-Fandino and N. Kimura unless otherwise indicated. Inundation map (Fig. 12.6) *developed by N. Kimura.*

References

Adityawan BM, Roh M, Tanaka H, Farid M (2012) The effect of river mouth morphological features on tsunami intrusion. Proceedings of the 8th international conference on disaster management (CD-ROM), pp 75–83. http://www.iiirr.ucalgary.ca/files/iiirr/A4-1_.pdf. Accessed Apr 2015

Atsumi M (2009) River Bureau, Ministry of Land, Infrastructure, Transport and Tourism, Japan. http://www.mlit.go.jp/river/basic_info/english/pdf/conf_05.pdf. Accessed Jan 2015

Biodiversity Center of Japan (2013) Impacts of the Great East Japan earthquake on the natural environment in Tohoku Coastal Regions. Nature Conservation Bureau; Ministry of the Environment, Japan. http://www.shiokaze.biodic.go.jp/PDF/pamphlet_en.pdf. Accessed Jan 2015

Clilverd H, Thompson J, Heppell K, Sayer C, Axmacher J (2015) Removal of river embankments and the modelled effects on river-floodplain hydrodynamics. Geophys Res Abstr 17, EGU2015-12974-1, 2015 EGU General Assembly 2015. http://meetingorganizer.copernicus.org/EGU2015/EGU2015-12974-1.pdf. Accesed Mar 2015

Dawson AG (1994) Geomorpholical effects of tsunami run-up and backwash. geomorphology 10(1–4):83–94. http://www.sciencedirect.com/science/article/pii/0169555X94900094. Accessed Jan 2015

Ellis JT, Sherman DJ (2015) Coastal and marine hazards, risk and disaster. Elsevier. https://books.google.com.mx/books. Accessed May 2015

Godou H (2010) River Basin management in Japan. Flood control measures-water resource management. Ministry of Land, Infrastructure, Transport and Tourism, Japan. http://www.mlit.go.jp/river/basic_info/english/pdf/conf_04-0.pdf. Accessed Apr 20145

Grzelak K, Szczuciński W, Kotwicki L, Sugawara D (2013) Ecological status of sandy beaches after tsunami events – insights from meiofauna investigations after the 2011 Tohoku-oki tsunami, Sendai, Japan. In: Kontar Y, Santiago V, Takahashi T (eds) Tsunami events and lessons learned; environmental and societal significance. Springer International Publishing AG. Cham, Switzerland

Hirano K (2013) Difficulties in post-tsunami reconstruction plan following Japan's 3.11 mega disaster: dilemma between protection and sustainability. J JSCE 1:1–11. https://www.jstage.jst.go.jp/article/journalofjsce/1/1/1_1/_pdf. Accessed Apr 2015

Imamura F, Anawat S (2012) Proceedings of the international symposium on engineering lessons learned from the 2011 Great East Japan earthquake, March 1–4, 2012, Tokyo, Japan. http://www.jaee.gr.jp/event/seminar2012/eqsympo/pdf/papers/118.pdf. Accessed Apr 2015

Inoue M, Kamogawa M (2012) The need to change the concept of water-related disaster prevention. Sci Technol Trends. http://www.nistep.go.jp/en/wp-content/uploads/NISTEP-STTO044-1E.pdf. Accessed Apr 2015

Iuchi K, Maly E, Johnson L (2014) Three years after a Mega-disaster: recovery policies, programs and implementation after the Great East Japan earthquake. In: Santiago-Fandiño V, Kontar Y, Kaneda Y (eds) Post-tsunami hazards: reconstruction and restoration. Springer, International Publishing AG. Cham, Switzerland

Japan Red Cross Society (2013) Evaluation of the JRCS and IFCR recovery and rehabilitation interventions after the Great East Japan earthquake and tsunami of 11 March 2011. Final report. September 2013. http://www.jrc.or.jp/vcms_lf/JPTsunamiEarthquakeEvaluation SEP2013.pdf. Accessed Apr 2015

Japan Water Forum (2012) http://www.waterforum.jp/eng/2011/Japan_Earthquake/matsushima/investigation/top_main_r.html#top. Accessed Dec 2014

JCOLD–Japan Commission on Large Dams (2009) Dams in Japan; past, present and future. CRC Press, Page 60. https://books.google.pl/books?id=If54C_G7.....ione%20typhoon%201948%20Japan&f=false. Accessed July 2015

Kamiyama M, Sugito M, and Kuse M (2012) Precursor of crustal movements before the 2011 Great East Japan earthquake. Proceedings of the international symposium on engineering lessons learned from the 2011 Great East Japan Earthquake, March 1–4, Tokyo, Japan. http://www.jaee.gr.jp/event/seminar2012/eqsympo/pdf/papers/8.pdf. Accessed Dec Jan 2014

Keller RK, Finlayson M, Humphries P (2015) Restoring and conserving nature in the anthropocene means changing our ideas of success. The conservation. September 7. https://theconversation.com/restoring-and-conserving-nature-in-the-anthropocene-means-changing-our-idea-of-success-42691. Accessed Sept 2015

Komatsu G, Goto K, Baker RV, Oguchi T, Hayakawa SY, Saito H, Pelletier JD, McGuire L, Iijima Y (2013) Effects of tsunami wave erosion on natural landscapes: examples from the 2011 Tohoku-oki tsunami. In: Kontar Y, Santiago V, Takahashi T (eds) Tsunami events and lessons learned; environmental and societal significance. Springer International Publishing AG. Cham; Switzerland

Matsuura K, Shinohara G, Nakae M (2009) Historical fish specimens collected from the Tohoku district by the Saito Ho-on Kai Museum of natural history. Bull Natl Mus Nat Sci Ser A 35(1):9–54. https://www.kahaku.go.jp/research/researcher/papers/211053.pdf. Accessed Mar 2015

MLIT–Ministry of Land, Infrastructure, Transport and Tourism, Japan (2014a) http://www.mlit.go.jp/river/toukei_chousa/kasen/jiten/nihon_kawa/map_82.html - in Japanese. Accessed Jan 2014

MLIT–Ministry of Land, Infrastructure, Transport and Tourism, Japan (2014b) White paper. creating and preserving a beautiful and healthy environment. Promoting counter measures against global warming, Section 1. https://www.mlit.go.jp/common/001063088.pdf. Accessed Apr 2015

Mori N, Takahashi T, Yasuda T, Yanagisawa H (2011) Survey of 2011 Tohoku earthquake tsunami inundation and run-up. Geophysical research letters, vol 38. http://onlinelibrary.wiley.com/store..... Accessed Mar 2015

MPG (Miyagi Prefectural Government) (2011) Overview of the rivers in jurisdiction area (in Japanese). http://www.pref.miyagi.jp/soshiki/ks-doboku/river001.html. Accessed May 2015

MPG (Miyagi Prefectural Government) (2014) Miyagi prefecture, river bank recovery (Minato River) (in Japanese). http://www.pref.miyagi.jp/uploaded/attachment/203358.pdf. Accessed May 2015

MT (Minamisanriku Town-Minatogawa –Utatsu-) (2015) Virtual museum (in Japanese). http://www.town.minamisanriku.miyagi.jp/museum/natural/article.php?p=34. Accessed May 2015

Nagasawa T, Tanaka H (2012) Study of structural damages with massive geomorphic change due to tsunami. International conference on disaster management/IIIRR annual international conference, Kumamoto, Japan. http://www.iiirr.ucalgary.ca/files/iiirr/A6-3_.pdf. Accessed Apr 2015

Nakagoshi N, Inoue M (2003) Rivers system in Japan from a landscape ecological aspect. J Environ Sci 15(2):160–166. Accessed 1 Dec 2014. www.jesc.ac.cn/jesc_cn/ch/reader/create_pdf.aspx?file_no

Nakamura K (2008) River restoration efforts in Japan. Overview and perspective. Public Works Research Institute, Japan. http://www.pwri.go.jp/eng/activity/pdf/reports/k.nakamura.080601.pdf. Accessed Feb 2015

Nakamura S (2015) Designing a deluge: history of the estimating method of flood designing. Japan technology and industry working group. Thinking about technology in history of modern Japan (19th–20th century). Technology in Modern East Asia History Workshop. June 11–12; 2015, UK. Daily Archives, April 27, 2015. http://japantech.hypotheses.org/date/2015/04/27. Accessed July 2015

Nakamura K, Tockner K, Amano K (2006) River and wetland restoration: lessons from Japan. Bioscience (Oxford). May 2006/vol. 56, no 5. http://bioscience.oxfordjournals.org/content/56/5/419.abstract. Accessed Apr 2015

NAS -National Academy of Science- (2012) Levees and the national flood insurance program. Report summary. Water Science and Technology Board. http://nas-sites.org/levees/report-summary/. Accessed Apr 2015

Oka F, Tsai P, Kimoto S, Kato R (2012) Damage pattern of river embankments due to the 2011 off the Pacific Coast of Tohoku earthquake and numerical modeling of the deformation of the river embankments with a clayey subsoil layer. Elsevier. Soils and Foundations 52(5):890–909, Oct 2012

Omachi T (1999) The river law with a commentary article. Legal framework for river and water management in Japan. Ministry of Construction; Infrastructure Development Institute-Japan. IDI water series no 4. http://www.idi.or.jp/library/pdf/RIVERE.PDF. Accessed 1 Dec 2014

Roh M, Bagus MA, Tanaka H (2014) Assessment of propagation characteristics for tsunami wave ascending river. Coastal engineering proceedings, no. 34. https://icce-ojs-tamu.tdl.org/icce/index.php/icce/article/view/7302/pdf_717. Accessed Apr 2015

S&TT (2012) Executive summary. Science and technology trends. Executive summary. http://www.nistep.go.jp/en/wp-content/uploads/NISTEP-STTO044-1E.pdf. Accessed Apr 2015

SAFOA and US Corps of Engineers (2015) American river watershed; draft environmental impact statement. Environmental impact report. http://yosemite.epa.gov/oeca/webeis.nsf/(eisdocs)/20150071/$file/american%20river%20common%20features%20draft%20eiseir.pdf?openelement. Accessed Apr 2015

Santiago-Fandiño V (2013) The tsunami and earthquake in Miyagi prefecture and sanriku coastline environmental, economic and social aspects 2011–2012. In: Kontar Y, Santiago V, Takahashi T (eds) Tsunami events and lessons learned; environmental and societal significance, 2013. Springer International Publishing AG. Cham, Switzerland

Santiago-Fandiño V, Kim MH (2015) Tsunami and environmental pollution: a note for the restoration process. In: Santiago-Fandiño V, Kontar Y, Kaneda Y (eds) Post-tsunami hazards; reconstruction and restoration. Springer International Publishing AG. Cham, Switzerland

Shadanhojin Tohoku Kensetsu Kyokai (2012) Tsunami Hisai Zengo no Kiroku. Kahoku Shimpo Publishing Centre, Sendai, Japan

Shi S (2003) Coastal tsunami geomorphological impacts and sedimentation processes: case studies of modern and prehistorical events. International conference on Estuaries and Coasts Nov 9–11, 2003, Hangzhou, China. http://www.irtces.org/pdf-hekou/021.pdf. Accessed Jan 2015

Shimazu H (2012) Geomorphic impacts of the 2011 tsunami on the lower reaches of the R. Natori, northeast Japan. Geophy Res Abstr 14:EGU2012-8986, 2012. http://meetingorganizer.copernicus.org/EGU2012/EGU2012-8986.pdf. Accessed Jan 2014

Shuto N, Fujima K (2009) A short history of tsunami research and countermeasures in Japan. Proc Jpn Acad Ser B Phys Biol Sci 85(8):267–275. http://www.ncbi.nlm.nih.gov/pmc/articles/PMC3621565/. Accessed Mar 2015

Taisei Corporation (2013) Development of cement-hardened material production technology for effective use of concrete rubble. 2013. JSCE Environmental Award Winning Technology.

http://www.jsce.or.jp/committee/concrete/e/newsletter/newsletter39/Newsletter39_files/data/jsce%20award/3.pdf. Accessed June 2015

Tanaka H, Kayane K, Bagus A M, Farid M (2013) The effect of bed slope to the tsunami intrusion into rivers. Coastal Dynamics. http://www.coastaldynamics2013.fr/pdf_files/154_Tanaka_Hitoshi.pdf. Accessed Apr 2015

Tanaka H, Kayane K, Adityawan M B, Roh M, Farid M (2014) Study on the relation of river morphology and tsunami propagation in rivers. Ocean Dyn 64:1319–1332. Springer. http://link.springer.com/article/10.1007/s10236-014-0749-y. Accessed Apr 2015

Tohoku Regional Bureau (MILT) (2015) Occurrence of the earthquake and tsunami approaching. Earthquake memorial museum. http://infra-archive311.jp/en/w01.html. Accessed Apr 2015

Tsuji Y, Satake K, Ishibe T, Harada T, Nishiyama A, Kusumoto S (2014) Tsunami heights along the Pacific Coast of Northern Honshu recorded from the 2011 and previous great earthquakes. Pure Appl Geophys 171(2014):3183–3215. http://download.springer.com/static/pdf/977/art%253A10.1007%252Fs00024-014-0779-x.pdf?auth66=1418721876_e3448ac2d7495cdc9f3d6cd6a3afaca7&ext=.pdf. Accessed Dec 2014

Udo K, Takeda Y, Takamura M, Mano A (2015) Serious erosion of the southern Sendai Coast due to the 2011Tohoku Earthquake Tsunami and its recovery process. In: Santiago V, Kontar Y, Kaneda Y (eds) Post tsunami hazards; reconstruction and restoration, 2013. Springer, International Publishing AG. Cham, Switzerland

Wikipedia-Japan (2014) http://ja.wikipedia.org/wiki/宮城県の二級水系一覧. Accessed Dec 2014

Yoshimura C, Omura T, Furumai H, Tockner K (2005) Present state of rivers and streams in Japan. River Res Appl 21:93–112. Accessed April 2015

Chapter 13
Tsunami Impacts on Eelgrass Beds and Acute Deterioration of Coastal Water Quality Due to the Damage of Sewage Treatment Plant in Matsushima Bay, Japan

Takashi Sakamaki, Youhei Sakurai, and Osamu Nishimura

Abstract The 2011 tsunami in Japan severely disturbed the benthic biota in Matsushima Bay (Miyagi Prefecture) via sediment erosion. Eelgrass beds markedly decreased from their original coverage of 2.2 km^2 in June 2007–0.02 km^2 in May 2012. Although our monitoring indicates that the eelgrass beds are slowly recovering, an analysis based on a Habitat Suitability Index model suggests that their recovery is restricted by deteriorated light conditions due to land subsidence resulting from the prior earthquake. Additionally, the tsunami severely damaged a sewage treatment plant adjacent to the bay, and this led to acute increases in chemical oxygen demand and coliform bacteria in the seawater. However, the water quality improved when the damaged sewage treatment plant started temporary operations, such as sedimentation, aeration, and chlorination, and recovered to pre-tsunami conditions in approximately 1.5 years. These results demonstrate that temporary treatment methods are important to minimize acute impacts of treatment facility damage on water environments.

Keywords Eelgrass • Habitat suitability • Land subsidence • Sanitary sewer • Temporary treatment

---

T. Sakamaki (✉)
International Research Institute of Disaster Science, Tohoku University, Sendai, Japan

Department of Civil and Environmental Engineering, Tohoku University, Sendai, Japan
e-mail: takashi.sakamaki.a5@tohoku.ac.jp

Y. Sakurai • O. Nishimura
Department of Civil and Environmental Engineering, Tohoku University, Sendai, Japan

V. Santiago-Fandiño et al. (eds.), *Tsunamis and Earthquakes in Coastal Environments*, Coastal Research Library 14, DOI 10.1007/978-3-319-28528-3_13

13.1 Introduction

The tsunami caused by the Great East Japan Earthquake in March 2011 resulted in tremendous infrastructure damage and drastically altered the landscapes of the northeastern coastal region of Japan. Land subsidence and tsunami-induced erosion resulted in severe damage to the biota of coastal habitats, such as sandy beaches, tidal flats, saltmarshes, seaweed beds, and eelgrass beds (e.g., Urabe et al. 2013; Hara 2014; Kanaya et al. 2014; Kawamura et al. 2014). Drastic changes in sediment properties (e.g., decreased muddy sediment cover and increased coarse sediment material cover) and benthic biota (e.g., decreased aquatic vegetation and increased opportunistic species) owing to the tsunami have been widely documented (e.g., Urabe et al. 2013; Hara 2014; Kanaya et al. 2014; Kawamura et al. 2014; personal communications with local fishermen). Many coastal habitats experienced erratic changes in their physical, chemical, and biological properties after the tsunami, and some coastal habitats are recovering (e.g., Hara 2014; Kanaya et al. 2014; Kawamura et al. 2014). In terms of ecosystem conservation, human-induced impediments for ecosystem recovery need to be identified and properly mitigated. Therefore, temporal variation in benthic biota needs to be monitored and carefully analyzed.

The tsunami affected coastal marine habitats not only directly, but also indirectly via alterations to material dynamics in watersheds and coastal systems. Owing to the tsunami, coastal populations decreased dramatically. This is expected to change the material loads from terrestrial to marine systems, and consequently may affect coastal marine ecosystem processes and structures. Furthermore, acute impacts of tsunami-induced alterations in material dynamics also need to be considered. While the tsunami inundation transported materials from the sea (e.g., seafloor sediments) to land, the tsunami backwash dragged tons of debris and materials from land to coastal waters (Lebreton and Borrero 2013; Goto et al. 2014). This included chemical feedstock, oil, and sewage from facilities damaged by the tsunami (e.g., Bird and Grossman 2011). In the later part of this chapter, we report a case of damage to a sewage treatment plant and the consequent impacts on coastal water environments.

13.2 Tsunami Damage and Recovery of Eelgrass Beds in Matsushima Bay

13.2.1 Loss of Eelgrass and Seaweed Beds

Matsushima Bay, which is located on the northeast coast of Japan (38°20′N, 141°05′E), had eelgrass beds of more than 10 km^2 before the 2011 tsunami (Ministry of the Environments, Japan 1992). The northeastern part of the bay had particularly high eelgrass bed coverage. The eelgrass beds of the bay were generally composed of *Zostera marina*. In addition, surveys in June 1996 and February 2001 estimated

Table 13.1 The estimated areas (km^2) of eelgrass beds and seaweed beds in Matsushima Bay (Nishimura 2013)

	May 2012 (after tsunami)	June 2007 (before tsunami)
Eelgrass beds	0.02	2.13
Seaweed beds	0.47	1.82

that eelgrass beds in the bay were approximately 10 and 13 km^2, respectively (Miyagi Prefecture 2006). However, a survey conducted in 2005 reported damaged eelgrass roots due to a chemical reduction in the sediment and increased hydrogen sulfide, and predicted future decreases in eelgrass beds (Ministry of the Environments, Japan 2009).

Based on onboard visual observations 8 months after the tsunami (November 2011), eelgrass beds in the bay were almost completely lost. In addition, seaweed beds on the rocky shores around the mouth of the bay decreased dramatically. Based on a comparison between data collected in June 2007 and May 2012, the tsunami resulted in decreases in eelgrass beds and seaweed beds (e.g., *Sargassum* spp.) by approximately 99 % and 74 %, respectively (Table 13.1). A field observation in June 2013 reported that the eelgrass coverage was slowly recovering and estimated that the eelgrass bed area was 0.068 km^2, which was three times greater than it was in 2012 (Fig. 13.1). However, this area was only approximately 3 % of the area before the tsunami. Compared to the delayed recovery of eelgrass beds in soft-bottom habitats, seaweeds in rocky shores, such as *Sargassum* spp. and *Eisenia* spp., generally recovered more rapidly.

13.2.2 Habitat Suitability Index Model to Assess Tsunami Impacts on Eelgrass

The great earthquake and tsunami of 2011 not only reduced eelgrass coverage, but also changed the physical properties of the environment, such as sediment properties and bed levels, which are tightly linked with the occurrence and growth of eelgrass (e.g., Goodman et al. 1995; Nielsen et al. 2002). We evaluated spatial and temporal variation in habitat suitability for eelgrass in the bay using the Habitat Suitability Index (HSI) (U.S. Fish and Wildlife Service 1980). In particular, we focused on how the tsunami affected environmental factors that are tightly linked with the survival and production of eelgrass, and evaluated whether eelgrass coverage had the potential to recover to pre-tsunami conditions. To develop an HSI model, we chose four environmental variables, water depth, water temperature, salinity, and sediment organic content, each of which potentially restricts the occurrence of eelgrass and is commonly measured in environmental monitoring programs. Using quantitative spatial distribution data for the four environmental variables in the bay, four Suitability Index (SI) values ranging between 0 (unsuitable) and 1 (optimum) were calculated based on the following criteria:

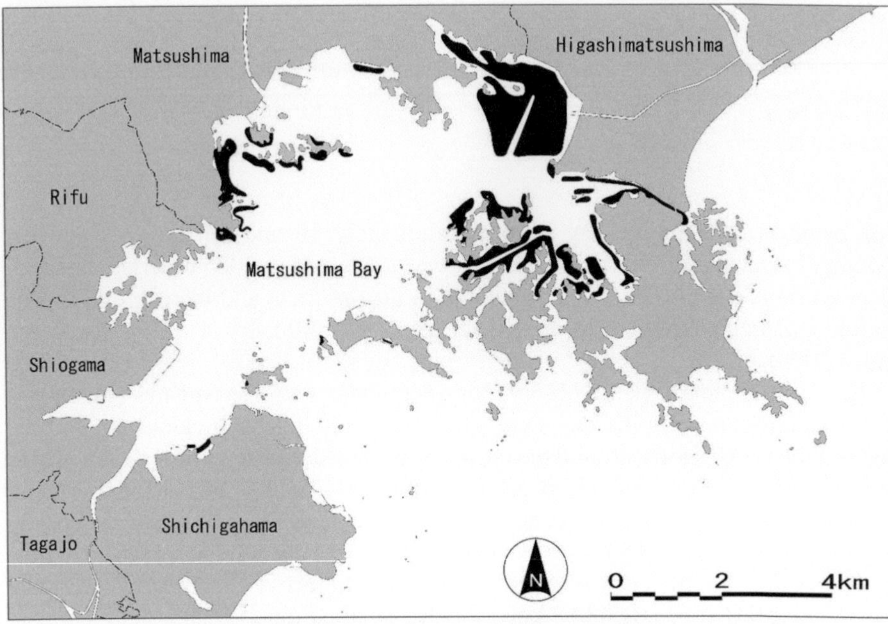

A. Before the tsunami (June 2007)

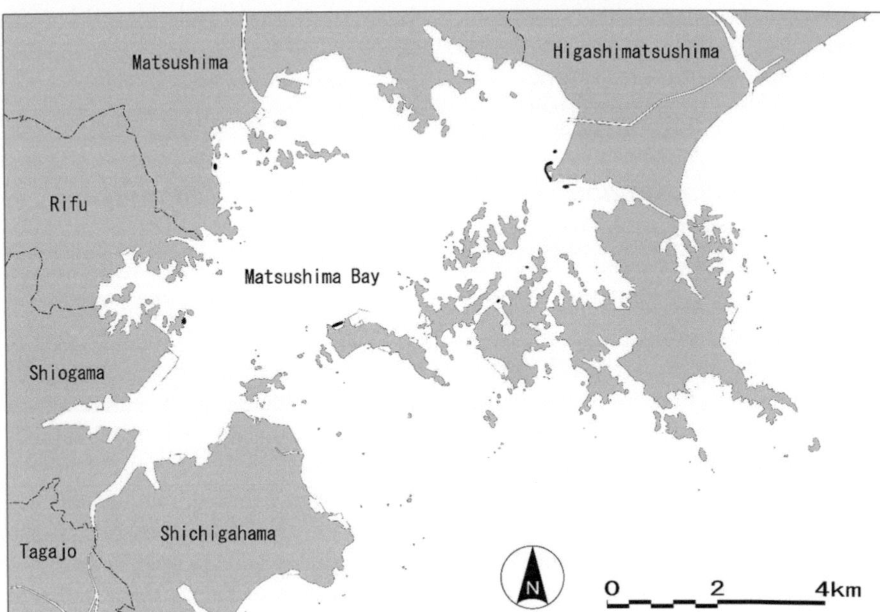

B. After the tsunami (June 2013)

Fig. 13.1 Spatial distributions of eelgrass beds in Matsushima Bay in June 2007 (pre-tsunami) and June 2013 (post-tsunami)

- Water depth

A decrease in water transparency owing to increased turbidity leads to decreased photosynthetic activity of eelgrass; accordingly, the photic environment is a crucial factor in determining eelgrass occurrence. Nielsen et al. (2002) showed that the depth limits of eelgrass are roughly equivalent to water transparency. Therefore, we used water depth as a proxy for the photic environment of the seabed, and calculated SI based on water transparency.

$SI_{WD} = 1$, if the water depth is less than the water transparency.
$SI_{WD} = 1 - $(water depth − transparency)/transparency, if the water depth is between the underwater transparency and double the transparency.
$SI_{WD} = 0$, if the water depth is greater than double the transparency.

- Water temperature

High water temperatures are associated with low ecological performance (e.g., photosynthesis, respiration, and growth) and increase eelgrass mortality (e.g., Nejrup and Pedersen 2008). Accounting for possible thermal stress on eelgrass during the summer, we applied the following criteria for water temperature as described in the guidelines for the restoration of eelgrass beds in coastal areas of Japan (Fisheries Agency of Japan and Marino-Forum 21 2007):

$SI_{WT} = 1$, if the water temperature is consistently <28 °C.
$SI_{WT} = 0$, if the water temperature reaches ≥28 °C during the summer.

- Salinity

The physiological performance of eelgrass declines in extraordinary high- or low-salinity conditions (e.g., van Katwijk et al. 1999; Nejrup and Pedersen 2008; Salo et al. 2014). We applied the salinity criteria for eelgrass survival indicated in the abovementioned guidelines for eelgrass restoration (Fisheries Agency of Japan and Marino-Forum 21 2007).

$SI_{SAL} = 1$, if the salinity is remains between 17‰ and 34‰.
$SI_{SAL} = 0$, if the salinity reaches <17‰ or >34‰.

- Chemical oxygen demand (COD) of sediment

Sediment with a high organic content causes physiological damage to eelgrass via hydrogen sulfide generation (e.g., Goodman et al. 1995). However, in our preliminary analysis, the SI_{COD} criteria provided by the abovementioned guidelines, that is, <10 mg-COD/g-sed (Fisheries Agency, Japan and Marino-Forum 21), were too strict and the HSI model evidently underestimated the suitability of habitats for eelgrass in the bay. Specifically, before the tsunami event, eelgrass covered areas with >30 mg-COD/g-sed in the bay (Miyagi Prefecture 2006).

Eelgrass is thought to prefer sandy sediment with relatively low mud and organic contents, but this is at least partly due to fine sediment resuspension and low water transparency in habitats characterized by sediment with high mud content (e.g., Vermaat et al. 1997; Nielsen et al. 2002). In Matsushima Bay, which is highly enclosed and generally has weak currents and waves, sediment resuspension occurs less frequently and a relatively preferable photic environment for eelgrass is maintained despite the predominance of muddy sediment. Hence, we applied the following SI_{COD} threshold, which was relatively moderate compared with the general criteria (i.e., $SI_{COD}=1$ at <10 mg-COD/g-sed), in our analysis.

$SI_{COD}=1$, if COD is <10 mg/g-sed.
$SI_{COD}=0.5$, if COD is between 10 mg/g-sed and 30 mg/g-sed.
$SI_{COD}=0$, if COD is ≥30 mg/g-sed.

Sediment grain size is another important factor for the occurrence of eelgrass. However, sediment grain size has a strong negative correlation with sediment organic content and was excluded from our analysis.

The SI_{WD} was calculated before the earthquake based on a 1/25,000 bathymetric chart of the bay provided by the Geospatial Information Authority of Japan in 1982. The other three SI indexes, SI_{WT}, SI_{SAL}, and SI_{COD}, were calculated based on environmental monitoring data provided by the local government (Miyagi Prefecture 2006). Finally, HSI was defined as the product of all four SI indexes, and can be interpreted as a quantitative site-specific description of habitat suitability for eelgrass.

Although our model was conservative with respect to threshold value and likely overestimated the area of suitable habitat for eelgrass in the bay, it was generally consistent with the spatial distribution of suitable habitats before the tsunami. In particular, the high HSI in the eastern part of the bay was consistent with the high coverage of eelgrass in the area pre-tsunami (Fig. 13.2a).

13.2.3 Changes in the Physical Environment and the Potential for Eelgrass Recovery

We assessed the potential for post-tsunami recovery of eelgrass in the bay using the abovementioned HSI model. In particular, we focused on the potential effect of the increased water depth associated with the earthquake-induced land subsidence, which was one of the most marked changes in the physical environment of the bay. The extent of land subsidence around the coastal area of Matsushima Bay was estimated at approximately 30–140 cm (based on field surveys by the Geospatial Information Authority of Japan, Matsushima Town Office, and Miyagi Prefecture).

The subsidence amounts differed among locations in Matsushima Bay. In addition, the land subsidence is slowly recovering since the Great East Japan Earthquake and Tsunami (Geospatial Information Authority of Japan 2014). However, in our analysis using the HSI model, we assumed for the sake of convenience that the water depth of the bay increased homogeneously by 50 cm. We also assumed that the other three environmental factors in the model did not change due to the earthquake or tsunami. As a result, our analysis suggests that the land subsidence

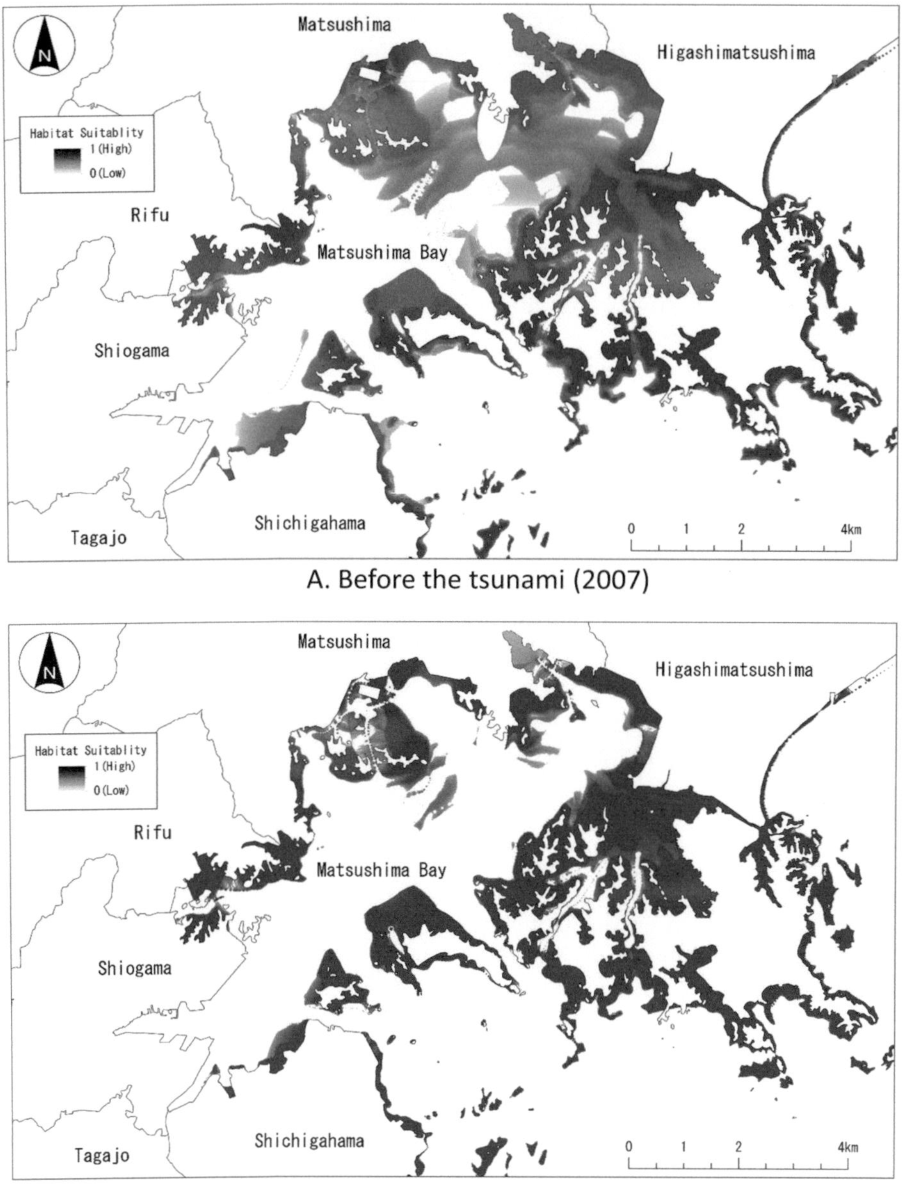

Fig. 13.2 Spatial distribution of the Habitat Suitability Index for eelgrass in Matsushima Bay. The Habitat Suitability was estimated for 2007 (pre-tsunami; *panel A*) and 2012 (post-tsunami; *panel B*)

potentially led to a marked decrease in the area suitable for eelgrass occurrence (Fig. 13.2b). Particularly in the central and eastern parts of the bay where eelgrass beds were dense before the tsunami, the increased water depth and disadvantageous photic environment likely hampered the current eelgrass bed recovery process.

In addition to land subsidence, other factors and/or interactions among multiple factors may also impede the recovery of eelgrass beds. Some previous studies examining shallow lakes have reported nonlinear relationships between nutrient concentration, phytoplankton density, water transparency, and submerged aquatic plant cover (Scheffer et al. 2001). In those models, when nutrient concentrations exceed a certain threshold, lakes abruptly transit from aquatic plant-predominant to phytoplankton-predominant systems. In particular, the drastic decrease in eelgrass in Matsushima Bay after the tsunami may have further altered properties of the physical environment that greatly affect the settlement and survival of eelgrass. For instance, aquatic plants generally reduce bed shear stress and suppress sediment resuspension in vegetated habitats (Hansen and Reidenbach 2012); accordingly, the loss of eelgrass is expected to increase sediment resuspension. This would further decrease light inputs to the seabed in the bay and negatively affect eelgrass recovery.

13.3 Acute Impact of Tsunami-Induced Damage to a Sewage Treatment Plant on Coastal Water Quality

13.3.1 Damage and Recovery of a Sewage Treatment Plant

The great earthquake and tsunami that occurred in March 2011 stopped operations of all sewage treatment plants located near the coast. Most of those stoppages in plant operations were due to power outages and damage to the facilities. One of the severely damaged sewage treatment plants was located in southwestern Matsushima Bay (Fig. 13.3). Immediately after the tsunami, facilities pumping sewage to the plant stopped operating, and sewage water spilled from the sewer. The untreated sewage flowed into an adjacent river and entered the bay. A plan for restoring this sewage plant consisting of the following four phases was announced soon after the plant stoppage:

Phase 1: Through June 2011. Repair facilities to receive sewage from sewer systems and operate temporary treatment processes by sedimentation and chlorination.

Phase 2: June 2011 – March 2012. Repair facilities to improve the quality of treated water via temporary aeration, and restart sludge treatment.

Phase 3: March 2012 – December 2012. Switch from temporary aeration to active sludge treatment.

Phase 4: December 2012 – end. Repair sludge incinerators and complete recovery of all facilities.

Fig. 13.3 (**a**) Aerial photograph of the Sen-en Sewage Treatment Center before the tsunami. The delineated area indicates the site of the treatment plant. (**b**) Sewage treatment facilities being submerged by the tsunami on March 11, 2011. (**c**) Active sludge tank filled with tsunami deposits. (**d**) Damaged anaerobic sludge digestion tank overturned by the tsunami. (**e**) Submerged pump facility. (**f**) Untreated sewage spilling from sewers near the treatment center due to the pump's failure (Source: Miyagi Prefectural Government (permission granted for publication))

13.3.2 Impacts on Water Quality

Regular marine water quality monitoring programs operated by local governments were discontinued owing to the devastation caused by the earthquake and tsunami. However, to account for the potential impacts of stoppages in sewage treatment plants on water environments, emergency surveys of water quality were conducted at several coastal sites in the region. In Matsushima Bay, the water quality was monitored in collaboration with the Miyagi Prefecture government and the author's research group at Tohoku University. The emergency survey in the bay indicated notable increases in the chemical oxygen demand (COD) after the tsunami at two monitoring stations, Minatobashi and Nishihama, which are located in the southwestern and central parts of the bay and are relatively close to the damaged sewage treatment plant (Fig. 13.4).

In Minatobashi, the increased COD recovered to the past average level within approximately 1–1.5 years. Likewise, in Nishihama, a notable decrease in COD was observed within 1 year, although it was slightly higher than the pre-tsunami level even 21 months after the tsunami. Other stations that are relatively far from the treatment plant continued to show usual COD levels after the tsunami. Furthermore, based on the emergency survey, the number of coliform bacteria in the seawater of the bay also markedly increased after the tsunami, and was estimated at 480,000 MPN/L in May 2011. However, it decreased to 9300 and 15,000 MPN/L in July and September 2011, respectively, close to the environmental standard established by the Japanese government, 10,000 MPN/L.

13.3.3 Importance of Temporary Treatment

The restoration of water quality observed after the tsunami was likely associated with the recovery of the damaged sewage treatment plant. In particular, the post-tsunami recoveries of COD and coliform bacteria after 2–6 months were associated with the first and second recovery phases of the damaged sewage treatment plant. During these two recovery phases, effluents of untreated sewage were blocked, and temporary treatments by sedimentation and chlorination were initiated. These temporary operations likely greatly contributed to improving the water quality in the bay.

13.4 Conclusions

The 2011 tsunami severely damaged sewage treatment plants in the coastal areas of northeastern Japan, leading to the acute deterioration of coastal water quality. However, our analysis of Matsushima Bay indicated that the deteriorated water quality owing to the damaged sewage treatment plant improved soon after temporary operations began. In addition, the complete recovery of water quality to pre-tsunami conditions was achieved as soon as plant operations completely recovered approximately 1.5 years after the tsunami. These water quality recovery processes observed in Matsushima Bay indicate the importance of not only reinforcing sewage treatment plant facilities, but also promptly establishing temporary operations to minimize the impacts of damaged treatment facilities on water environments.

The tsunami severely disturbed the benthic biota of the coastal marine areas of northeastern Japan. Eelgrass beds of Matsushima Bay decreased drastically after the tsunami, and are still slowly recovering. Their recovery is possibly hampered by deteriorated light conditions owing to land subsidence resulting from the earth-

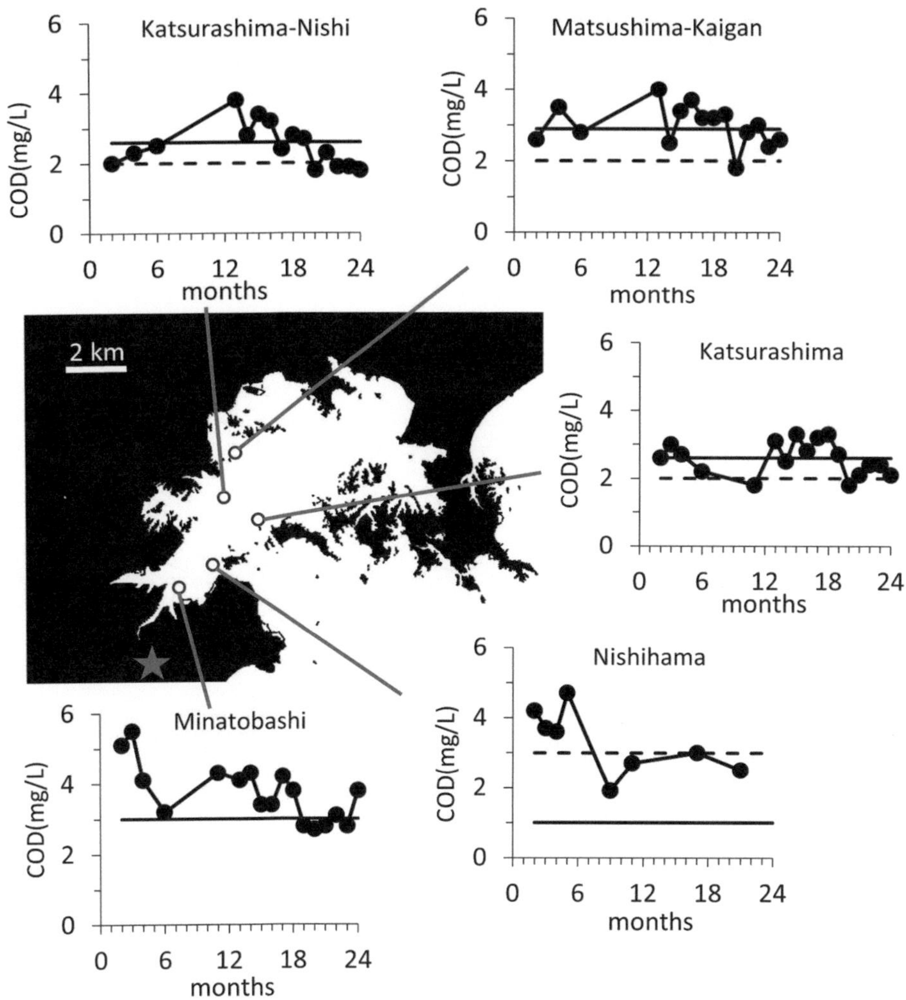

Fig. 13.4 COD temporal variation at five stations in Matsushima Bay after the tsunami (*oscillating line*). *Solid line* indicates COD average between 2007 and 2010 in four stations. *Dotted line* indicates the environmental water quality standard set established by the government which varies in accordance with the water usage and the environmental characteristics of the site i.e. 2, 3, or 8 mg/L (e.g., the water quality standard for Minatobashi is set at 8 mg/L). The *white star* indicates the location of the damaged sewage treatment plant. (Data source: The graphs result from the combination of data obtained by the author's research group (Nishimura 2013) and the one provided by Miyagi Prefectural Government)

quake. The loss of eelgrass in the bay may also increase sediment resuspension and decrease light inputs to the seabed, negatively affecting the recovery of eelgrass in the bay. In addition, human populations and land-use adjacent to the bay changed dramatically after the tsunami. These changes may alter nutrient loads in watersheds, and consequently affect the competition between eelgrass and phytoplankton in the bay. Possible human-induced impediments to the recovery of eelgrass beds in the bay should be carefully examined via environmental monitoring and properly mitigated. Furthermore, eelgrass beds play important ecological roles, including as nurseries and refuges for various marine organisms; therefore, the ecological effects of eelgrass loss in the bay need to be carefully monitored.

References

Bird WA, Grossman E (2011) Chemical aftermath: contamination and cleanup following the Tohoku earthquake and tsunami. Environ Health Perspect 119:290–301

Fisheries Agency of Japan and Marino-Forum 21 (2007) Guideline for restoration of eelgrass beds (in Japanese)

Geospatial Information Authority of Japan (2014) Crustal movements in the Tohoku District. Report from the Coordinating Committee for Earthquake Prediction, Japan, vol 92

Goodman JL, Moore KA, Dennison WC (1995) Photosynthetic responses of eelgrass (*Zostera marina* L.) to light and sediment sulfide in a shallow barrier island lagoon. Aquat Bot 50:37–47

Goto K, Hashimoto K, Sugawara D, Yanagisawa H, Abe T (2014) Spatial thickness variability of the 2011 Tohoku-oki tsunami deposits along the coastline of Sendai Bay. Mar Geol 358:38–48

Hansen JCR, Reidenbach MA (2012) Wave and tidally driven flows in eelgrass beds and their effect on sediment suspension. Mar Ecol Prog Ser 448:271–287

Hara K (2014) Damage to coastal vegetation due to the 2011 tsunami in northeast Japan and subsequent restoration process: analyses using remotely sensed data. Global Environ Res 18:27–34

Kanaya G, Maki H, Suzuki T, Sato-Okoshi W, Kikuchi E (2014) Tsunami-induced changes in a shallow brackish lagoon ecosystem (Gamo Lagoon) facing Sendai Bay, Japan. Global Environ Res 18(1):35–46

Katwijk V, Schmitz MM, Gasseling GHW, Van Avesaath AP, Van PH (1999) Effects of salinity and nutrient load and their interaction on *Zostera marina*. Mar Ecol Prog Ser 190:155–165

Kawamura T, Takami H, Hayakawa J, Won N, Muraoka D, Kurita Y (2014) Changes in abalone and sea urchin populations in rocky reef ecosystems on the Sanriku Coast damaged by the massive tsunami and other environmental changes associated with the Great East Japan Earthquake in 2011. Global Environ Res 18:47–56

Lebreton LCM, Borrero JC (2013) Modeling the transport and accumulation floating debris generated by the 11 March 2011 Tohoku tsunami. Mar Pollut Bull 66:53–58

Ministry of the Environment, Japan (1992) Reports of the 4th National Survey of Natural Environment (in Japanese)

Ministry of the Environment, Japan (2009) Reports of the 7th National Survey of Natural Environment (in Japanese)

Miyagi Prefecture (2006) Reports of the committee of refreshment Project of Matsushima Bay (in Japanese)

Nejrup LB, Pedersen MF (2008) Effects of salinity and water temperature on the ecological performance of Zostera marina. Aquat Bot 88:239–246

Nielsen SL, Sand-Jensen K, Borum J, Geertz-Hansen O (2002) Depth colonization of eelgrass (Zostera marina) and macroalgae as determined by water transparency in Danish coastal waters. Estuaries 25:1025–1032

Nishimura O (2013) Effects of the earthquake event of 2011 on water environments of Matsushima Bay. J Jpn Soc Water Environ 36:49–52 (in Japanese)

Salo T, Pedersen MF, Boström C (2014) Population specific salinity tolerance in eelgrass (Zostera marina). J Exp Mar Biol Ecol 461:425–429

Scheffer M, Carpenter S, Foley JA, Folke C, Walker B (2001) Catastrophic shifts in ecosystems. Nature 413:591–596

U.S. Fish & Wildlife Service (1980) Habitat evaluation procedure handbook. Division of Ecological Services, Washington, DC

Urabe J, Suzuki T, Nishita T, Makino W (2013) Immediate ecological impacts of the 2011 Tohoku earthquake tsunami on intertidal flat communities. PLoS ONE 8:e62779

Vermaat JE, Agawin NSR, Fortes MD, Uri JS, Duarte CM, Marba N, Enriquez S, van Vierssen W (1997) The capacity of seagrasses to survive increased turbidity and siltation: the significance of growth form and light use. AMBIO 26:499–504

Chapter 14
Effects of the Great East Japan Tsunami on Fish Populations and Ecosystem Recovery. The Natori River; Northeastern Japan

Kinuko Ito, Ayu Katayama, Kazunori Shizuka, and Norihiro Monna

Abstract The tsunami following the Great East Japan Earthquake in March 2011 resulted in significant ground subsidence and deposition of rubble and mud in the Natori River, near the city of Sendai (Miyagi Prefecture), damaging its brackish water ecosystem and fishing grounds. There was a direct impact in the form of annihilation of animal and plant life and disturbance of the habitats throughout. Also, a wedge of seawater ran far upstream, and ground subsidence changed the pattern of tidal flow in the river. Brackish water ecosystems such as that near the mouth of the Natori River are important as nurseries for juvenile fishes and as a fishing ground for bivalves such as clams. The populations of both of these kinds of organism declined drastically as result of the tsunami. The catch per unit effort of ayu fish (sweetfish; *Plecoglossus altivelis altivelis*) in 2011 was the lowest recorded for the past 5 years, and the population hatching date composition showed a marked absence of early-hatched individuals. In contrast, the residual upstream ayu fish population seems to have grown successfully and reproduced despite the effects of the tsunami: 1 year after the tsunami occurred, the downstream ayu fish population had recovered to the same level as before the event. However, the population of the brackish-water clam, *Corbicula japonica*, only showed recovery 2 years after the disaster as its habitat has drastically shifted due to movement of the brackish water zone about 1 km upstream. The studies reported here show that the impact of the earthquake and tsunami on pelagic fish and benthic bivalves seems to have been quite different, as in the former recovery was rapid, while in the latter it took much longer. Many other fish species also returned to normal levels within a year, such as stone flounder (*Kareius bicoloratus*), goby (*Acanthogobius lactipes*), icefish (*Salangichthys microdon*) and black porgy (*Acanthopagrus schlegelii*). The food web structure appears to be slightly different from past years, but the results show that, in general, fish communities are able to recover rapidly from disturbances even as drastic as an unusually large tsunami.

K. Ito (✉) • A. Katayama • K. Shizuka • N. Monna
Graduate School of Agricultural Science, Tohoku University,
1-1, Tsutsumidori Amamiya machi, Aobaku, Sendai, Japan
e-mail: kinuko.ito.c6@tohoku.ac.jp; itokn@bios.tohoku.ac.jp

V. Santiago-Fandiño et al. (eds.), *Tsunamis and Earthquakes in Coastal Environments*, Coastal Research Library 14, DOI 10.1007/978-3-319-28528-3_14

Keywords Natori River • Fish population • Ecosystem • Effects • Recovery

14.1 Introduction

Scientists failed to predict the huge earthquake off the Pacific coast of Tohoku on 11 March, 2011. The subsequent tsunami struck not only human life but also organisms living in the water. Substantial changes occurred in the local environment, and many areas suffered from significant ground subsidence and deposition of large amounts of debris, including rubble from coastal constructions destroyed and dragged seaward by the receding tsunami.

Miyagi Prefecture is the second largest fishery landing region in Japan and as a result of the tsunami this fishery was heavily affected: many ships were lost; ports and jetties were destroyed (Watanabe et al. 2013; Fisheries Research Agency of Japan 2013); and important fish resources such as beds of abalone and sea urchin were damaged (Takami et al. 2013).

Sampling sites for the present study are at the mouth of the Natori River, which is an important fishing ground both for bivalves and fishes. The river mouth was severely eroded by the tsunami, and the water depth is now significantly deeper than before the tsunami (Tanaka et al. 2013). Various fish species live in brackish water areas, which are very important for the maintenance of fishery resources, particularly the juvenile stages of fishes and bivalves. Juveniles of amphidromous fish such as salmon and sweetfish have their juvenile stages during spring: salmon juveniles swim out to the ocean; while ayu fish run upstream once the water temperature exceeds 10 °C.

Surveys of biological production systems in this area were begun over 10 years ago in order to elucidate the functions of brackish areas within the context of environmental conservation and fishery production. In addition to focusing on the brackish water nursery for local fishery resources, the data accumulated enable comparisons to be made concerning the effects of the tsunami and subsequent recovery. Studies have been conducted independently on ayu fish, clam, and fish communities, the results of which are reviewed and discussed in this chapter in relation to the effects of the tsunami.

14.2 Methods and Approach

14.2.1 Location and Study Sites

The Natori River in Miyagi Prefecture is listed as a first class river according to the River Act of Japan (Ministry of Land, Infrastructure,Transport and Tourism (2013)). It is 55 km in length with a basin area of 939 km^2 and has 13 branches, one of which

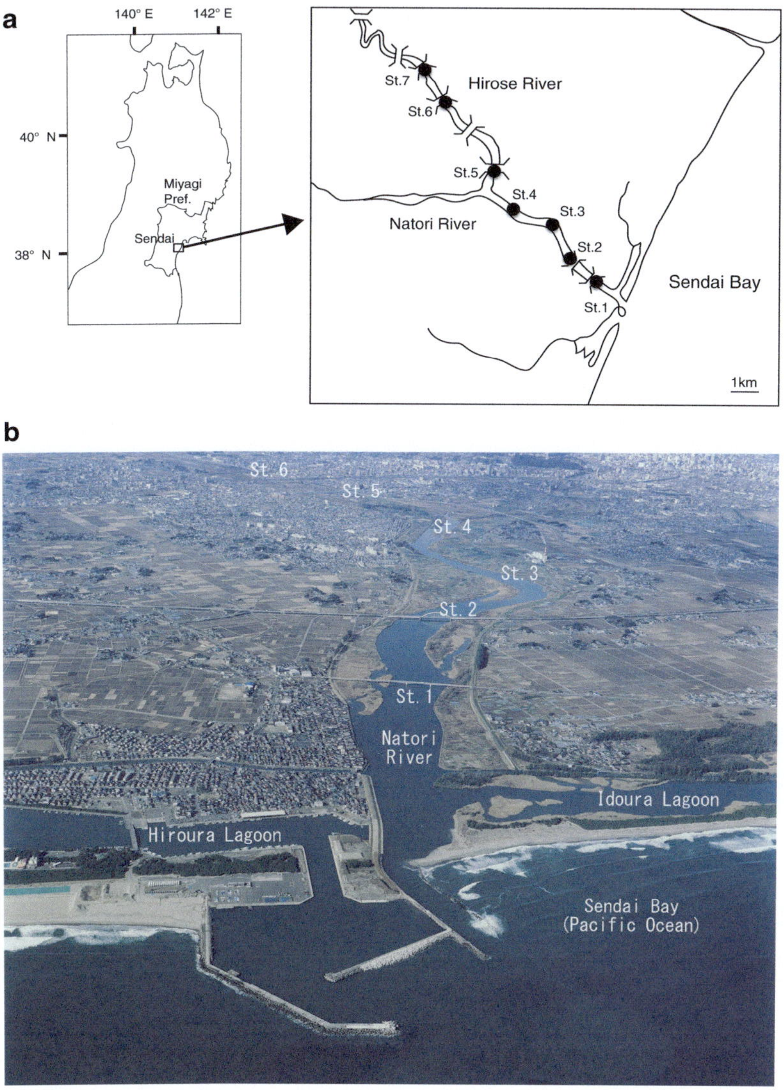

Fig. 14.1 (**a**) Map of the study site and location of sampling stations. (**b**). Aerial view of study sites on 8th December 2005, before the tsunami (From Sendai River and National Highway Office)

is the Hirose River, which passes through the city of Sendai. Figure 14.1a shows the location of the stations at which samples are taken regularly. Stations 1–7 are positioned upstream along the Hirose branch at 1.5, 2, 4, 6, 8, 12 and 14 km from Sendai Bay. Station 1 is a fish community study site; Stations 1–4 are clam study sites; and Stations 5–7 are ayu fish study sites. Figure 14.1b shows an aerial view of the study sites on 8th December 2005, before the tsunami.

14.2.2 Survey of Physical Environmental Conditions

Water temperature and salinity were measured in the field whenever sampling was conducted. To obtain accurate diurnal changes in salinity, data-loggers (YSI600OMS) have been installed since 2011 Bottom sediments are also collected and analyzed at regular intervals.

14.2.3 Collection of Organisms and Measurements

Ayu fish (sweetfish, *Plecoglossus altivelis altivelis*) were collected using a casting net at three stations in 2011 and 2012. Catch per unit effort (CPUE) was calculated as the mean number of individuals caught in 10 net casts. Hatch date composition of these individuals was analyzed based on daily otolith rings, and gonadosomatic index (GSI) was calculated from total wet weight and gonad wet weight.

Other fish were collected by beach seine at Station1 from April 2011 to March 2012 and subsequently identified to species level, in order to assess and analyze the character of the fish community. Measurements were taken of length and wet weight. In order to analyze food source utilization, stomach contents were analyzed, and carbon and nitrogen stable isotope measurements were taken from muscle samples.

Clams (the brackish-water clam, *Corbicula japonica*) were collected at Stations 1–3 from May 2009 to April 2010, with a further station (4) added after the earthquake (data for 2011–2013).

14.2.4 Food Source and Food Web Analysis Using Stable Isotope Ratios

Stable isotope ratios for carbon ($\delta^{13}C$) and nitrogen ($\delta^{15}N$) have been widely applied to determine the food source of different species in order to analyze the structure of the food web (Deniro and Epstein 1978, 1981). The isotope ratio for carbon ($\delta^{13}C$) is characteristic of primary food sources that form the primary producer level in food webs: terrestrial plants, phytoplankton and sea grasses. The relationship between animals and their diet (food source) is roughly 0–1‰ $\delta^{13}C$ enrichment per trophic level (Fry and Sherr 1984). Trophic level is indicated by $\delta^{15}N$ enrichment, which rises at a rate of about 3.4‰ per trophic level (Minagawa and Wada 1984). There are significant variations in δ^{15} N of particulate organic nitrogen (PON) observed in different sampling locations around the world. Suspended particulate matter in the euphotic zone is typically lighter in the ^{15} N isotope in oligotrophic compared to eutrophic seas (Saino and Hattori 1980; Minagawa and Wada 1984). The present study uses plots of δ^{13} C and δ^{15} N of primary producers (planktonic diatoms and benthic diatoms), including data from field incubation experiments at the Natori River estuary in 1998 (Ito 2002).

14.3 Results

14.3.1 Effects on the Ayu Fish Population

Ayu fish is a commercially important resource, a fish with a longevity of about 1 year. Figure 14.2 is an overview of the ayu fish life cycle, showing that ayu fish juveniles were hit by the tsunami just as they were preparing to move upstream. In 2011, just after the tsunami, the CPUE at Station 5 was 1.1–4.5 individuals per cast during early May to late June, the lowest of the past 5 years. In 2012, the CPUE had recovered to the same level as before the tsunami (Fig. 14.3a). However, the CPUE at Station 6 remained below 1 fish per cast even after late May. Ayu fish running upstream can usually be observed from early May, but they were apparently delayed in 2011, with the CPUE recovering in 2012 (Fig. 14.3b). The hatching date composition shows a marked absence of fish that had hatched before late October 2010 (Fig. 14.4).

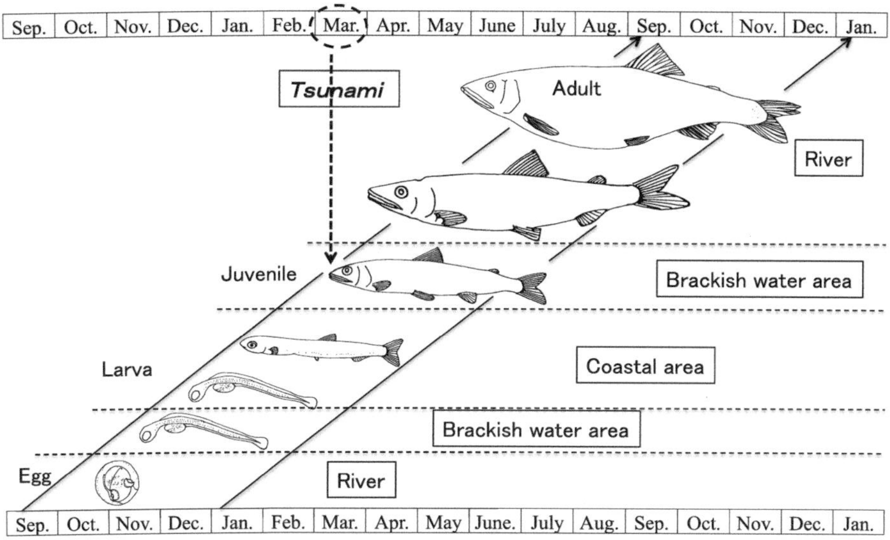

Fig. 14.2 Outline of developmental stages and habitats of Ayu fish (From Shizuka 2013) and tsunami hit date

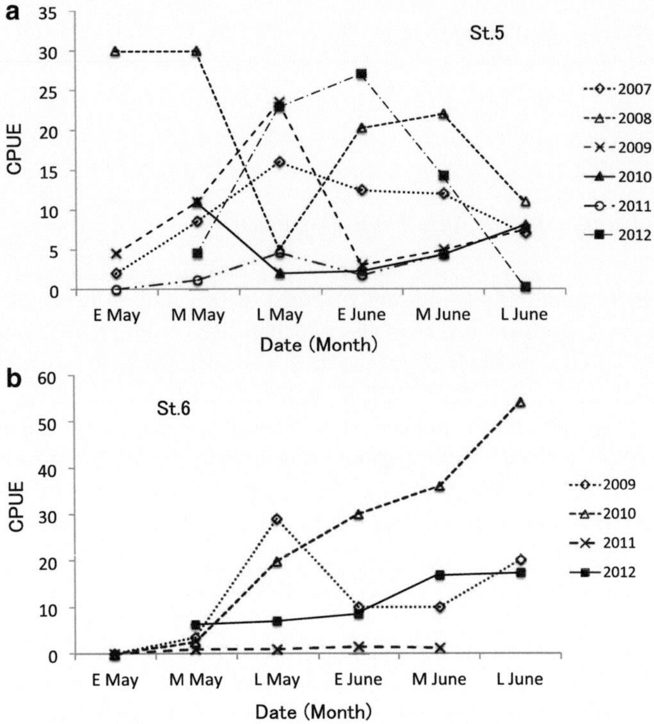

Fig. 14.3 Catches per unit effort of ayu fish running upstream, sampled (**a**) at St. 5 from 2007 to 2012, and (**b**) at St. 6 from 2009 to 2012

14.3.1.1 Maturation and Reproduction

Female gonads developed as usual, with GSI similar to those prior to the tsunami (Fig. 14.5). Regardless of the negative influence of the tsunami on the ayu fish population running upstream, the ayu fish river population showed little change in subsequent growth and reproduction compared with previous years. Research in 2012 denoted a large number of fish running upstream (described above), suggesting that the ayu fish population suffered little disturbance. Large variations in body size and hatching date in the ayu fish population in the Natori and Hirose rivers (Shizuka 2013) may have been a risk dispersing factor enabling the population to resist the effects of the tsunami. It is known that juvenile ayu fish hatching very late, in December or even January, can only survive against high sea surface temperatures in the Shimanto Estuary and adjacent coastal waters in Kochi Prefecture, Shikoku (Takahashi 2005).

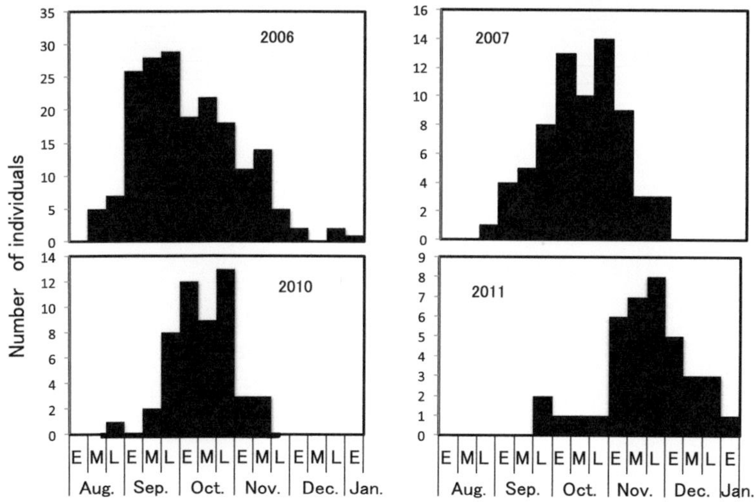

Fig. 14.4 Hatch date composition of Ayu fish sampled in 2006, 2007, 2010 and 2011 (From Shizuka et al. 2012)

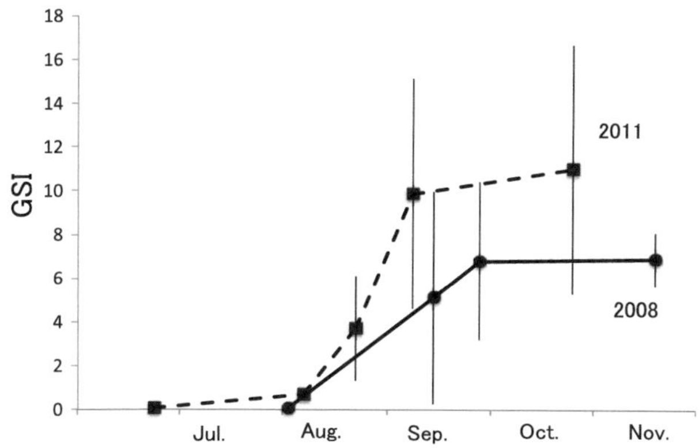

Fig. 14.5 Temporal changes in Gonadosomatic index (GSI) of Ayu fish sampled in 2008 and 2011 (From Shizuka et al. 2012)

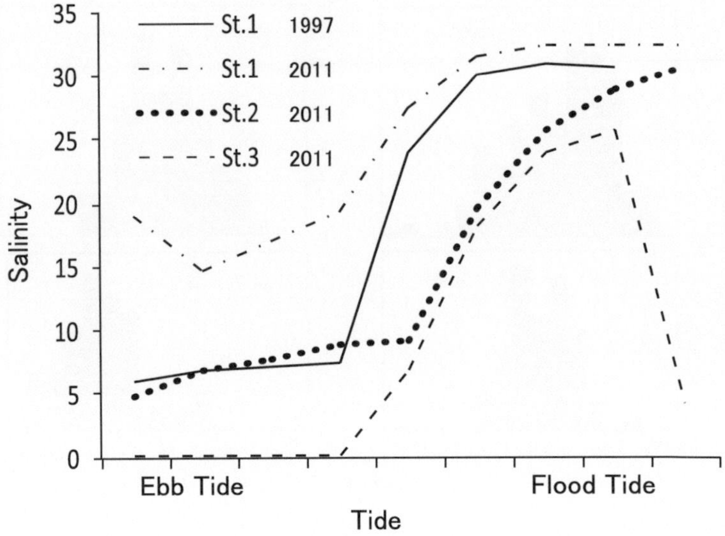

Fig. 14.6 Diurnal variations in salinity of water at the seabed for each station (From Katayama and Ito 2012)

14.3.2 Effects on the Population of Corbicula japonica

14.3.2.1 Distribution of Abundance

Corbicula japonica (Yamatoshijimi) is a commercially valuable bivalve of Japanese inland fisheries. However, the tsunami following the Great East Japan Earthquake drastically changed the brackish water area of the Natori River, which is the habitat favored by *C. japonica*. Figure 14.6 shows diurnal variations in the salinity of the water at the seabed for each station. The salinity fluctuated with the tide from 5 psu (at ebb tide) to 30 psu (flood tide) at Station1 before the earthquake (data from 1977) but shifted to higher salinity conditions between 15 psu (ebb tide) and 32 psu (flood tide) after the tsunami. The post-tsunami range of salinity at Station 2 is similar to that recorded at Station1 before the tsunami. These measurements indicate that the brackish area has extended upstream, presumably caused by ground subsidence in this area (around 0.25 m; Geospatial Information Authority of Japan 2015).

Figure 14.7 shows changes in the abundance of clams of commercial size. Abundance at all stations was ≥10 ind. m^{-2} before the earthquake, and all showed a decrease after the tsunami. At Station1, abundance was 3.7 ind. m^{-2} in July 2011, decreasing further to just 1 ind. m^{-2} since August of that year. The salinity tolerance range of *C. japonica* is 0–22 psu (Ishida and Ishii 1972), so this result clearly reflects the shifting of the brackish area upstream, depressing the abundance of *C. japonica*

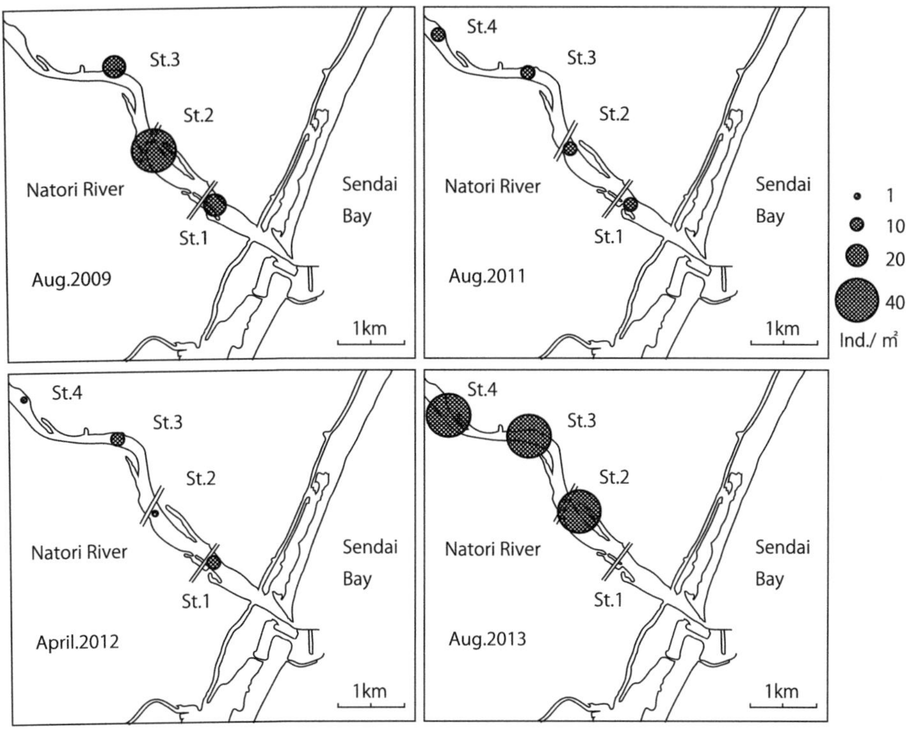

Fig. 14.7 Abundance of *Corbicula japonica* of commercial size (larger than 15 mm in shell length) sampled in 2009, 2011, 2012, 2013. *Circle* size indicates the density of individuals m^{-2} (Data from Katayama and Ito 2014)

at Station1, just 1.5 km from the river mouth. After 2 years, clam abundance has returned to approximately the same levels as before the tsunami, but with an upstream shift in the *C. japonica* habitat of about 1 km. Many juvenile *C. japonica* have been observed at Stations2, 3 and 4.

14.3.2.2 Changes in the Food of *C. japonica*

Figure 14.8 shows the mean $\delta^{13}C$ for *C. japonica* at Stations 1–3 before the tsunami and seasonal variation in $\delta^{13}C$ at Stations 1–4 after. In general, animal $\delta^{13}C$ reflects that of the diet, allowing estimation of the food source (Fry and Sherr 1984; Minagawa and Wada 1984). Organic input to estuarine systems can include terrestrial plant material ($\delta^{13}C$ around −29‰), benthic microalgae ($\delta^{13}C$ around −16‰) and phytoplankton (around −20‰) (France 1995; Ito 2002). All stations had shown stable values for $\delta^{13}C$ throughout the year until the tsunami. Previous reports on food sources for upstream *C. japonica* indicated a high contribution of terrestrial organic carbon (Katayama et al. 2013). However, after the tsunami, upstream $\delta^{13}C$

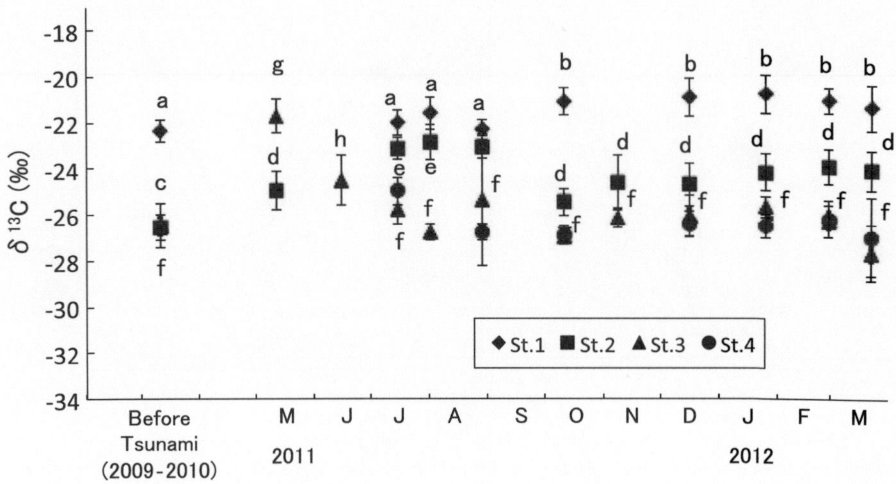

Fig. 14.8 Temporal changes of carbon stable isotope ratio ($\delta^{13}C$) in *Corbicula japonica*. Different letters indicate significant differences within each station at the 5 % level (From Katayama and Ito 2014)

values changed markedly. At 2 km upstream, a pre-tsunami value of −26.5‰ approached −23‰ in the 6 months after the tsunami, returning to around −26‰ by the 7th month. At 4 km upstream, the value rose from −26 to −22‰ immediately after the tsunami disturbance, gradually returning to −26‰ within 4 months. This is interpreted as an effect of the draining of terrestrial organic matter into the river following inundation by, and subsequent receding of, the tsunami water. The temporary increase of $\delta^{13}C$ of *C. japonica* in both areas probably reflects a relative increase in the contribution of microalgae utilizing the large amount of deposited organic matter.

14.3.3 Effects on River Function as a Nursery Ground for Fishes

14.3.3.1 Species Composition of Fishes

A total of 2162 individuals of 21 species of fish were obtained from sampling (Fig. 14.9 and Table 14.1). Three groups of life style patterns were apparent: resident, seasonal migrant and occasional.

Typical resident types *(R)* include the *Salangichthys microdon* (Shirauo), and gobies such as *Acanthogobius lactipes (Ashishirohaze)*. Many seasonal migrants *(S)* were encountered, as follows. In spring, *Plecoglossus altivelis altivelis* (Ayu), *Oncorhynchus keta* (Sake), *Kareius bicoloratus* (Ishigarei); in summer, *Tribolodon*

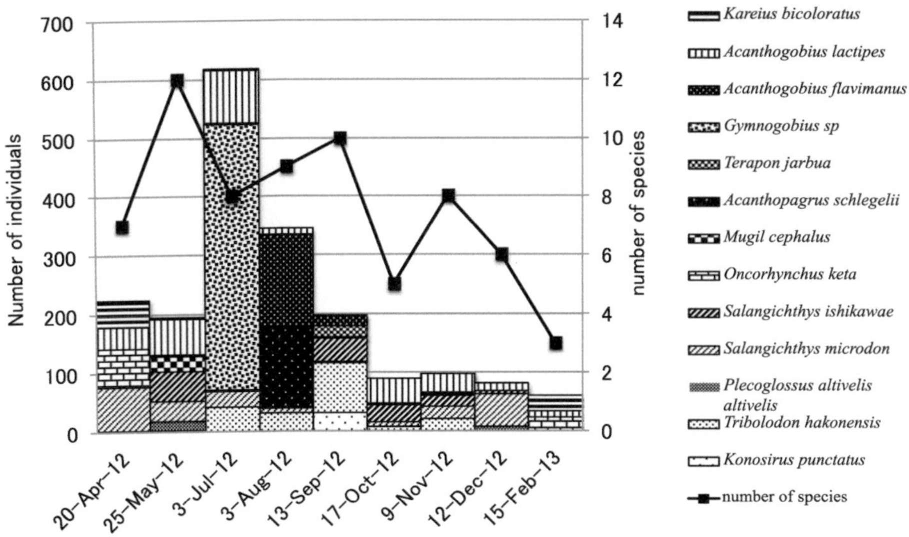

Fig. 14.9 Seasonal change of fish species composition at the mouth of the Natori River from April 2012 to February 2013

hakonensis (Ugui), *Gymnogobius* sp.1 (Sumiukigori), *Mugil cephalus* (Bora), *Acanthopagrus schlegelii* (Kurodai), and *Takifugu niphobles* (Kusafugu); in autumn and winter, *Salangichthys ishikawae* (Ishikawa-shirauo) and *Tridentiger obscurus* (Chichibu). Occasional species *(O)* included *Konosirus punctatus* (Konoshiro), *Pholis crassispina* (Takegimpo) and *Repomucenus valenciennei* (Hatatatenumeri). Many of these fishes were juveniles, with only three species present as adult fishes: *Salangichthys microdon* (Shirauo), *Leucopsarion petersii* (Shirouo) and *Acanthogobius lactipes* (Ashishirohaze). In terms of the number of species, the highest occurrence was in spring to summer. This is similar to the fish composition reported previously for this area (Ito and Kakegawa 2007; Honda et al. 1997).

14.3.3.2 Characteristics of Food Chains Based on Stable Isotope Ratios

The basic food web structure is similar to that reported previously (Ito and Kakegawa 2007). Few fishes were found to occupy higher trophic levels in this area. Most were around trophic level 3 or 4. The food chains show three fundamental pathways from phytoplankton, benthic diatoms, and detritus, the latter including terrestrial plant material (Fig. 14.10). However, relations among the fishes were not clear-cut, most species apparently depending upon a wide range of food items. There are slight post-tsunami differences in contributions from benthic and pelagic food sources, with $\delta^{13}C$ values for the same species shifted lower by about 1‰ compared with a previous report (Ito and Kakegawa 2007). Most species showed a shift to utilizing phytoplankton as a food source, and a lesser degree of dependence on benthic

Table 14.1 Fish species sampled at the Natori River mouth (April 2012 to February 2013)

Family	Scientific name	English	Japanese	Type
Clupeomorpha	*Konosirus punctatus*	Gizzard shad	Konoshiro	O
Cyprinidae	*Tribolodon hakonensis*	Japanese dace	Ugui	S
Plecoglassidae	*Plecoglossus altivelis altivelis*	Ayu	Ayu	S
Salangidae	*Salangichthys microdon*	Icefish	Shirauo	R
Salangidae	*Salangichthys ishikawae*	Ishikawa icefish	Ishikawa-shirauo	S
Salmonidae	*Oncorhynchus keta*	Chum salmon	Sake	S
Mugilidae	*Mugil cephalus cephalus*	Mullet	Bora	S
Platycephalidae	*Platycephalus indicus*	Flathead	Magochi	S
Sparidae	*Acanthopagrus schlegelii*	Black porgy	Kurodai	S
Teraponidae	*Terapon jarbua*	Crescent perch	Kotohiki	S
Pholidae	*Pholis crassispina*	Blenny	Takeginpo	O
Callionymidae	*Repomucenus valenciennei*	Whipfin dragonet	Hatatatenumeri	O
Gobiidae	*Leucopsarion petersii*	Ice goby	Shirouo	S
Gobiidae	*Gymnogobius* sp.1	Floating goby	Sumiukigori	S
Gobiidae	*Gymnogobius castaneus*		Biringo	S
Gobiidae	*Acanthogobius flavimanus*	Common Japanese goby	Mahaze	S
Gobiidae	*Acanthogobius lactipes*	Whitelimbed goby	Ashishirohaze	S
Gobiidae	*Tridentiger obscurus*	Threadfin goby	Chichibu	R
Pleuronectidae	*Platichthys stellatus*	Starry flounder	Numagarei	S
Pleuronectidae	*Kareius bicoloratus*	Stone flounder	Ishigarei	S
Tetraodontidae	*Takifugu niphobles*	Puffer	Kusafugu	S

Type of life pattern: *O occasional, S seasonal, R resident*

organisms. Presumably, this reflects decreases in the availability of benthic animals and benthic microalgae (such as epipsammic diatoms) following the tsunami.

14.4 Discussion

14.4.1 Differences in Effects of the Tsunami Between Pelagic and Benthic Communities

In the present series of studies, important effects of the environment were seen on the organisms involved in the dynamic processes of local biological production. The effects observed differed according to particular environmental niches. The ayu fish population recovered quickly during the year following the tsunami, and the residual group reproduced normally. Juveniles of the residual group in other areas are recruiting normally (Shizuka 2013).

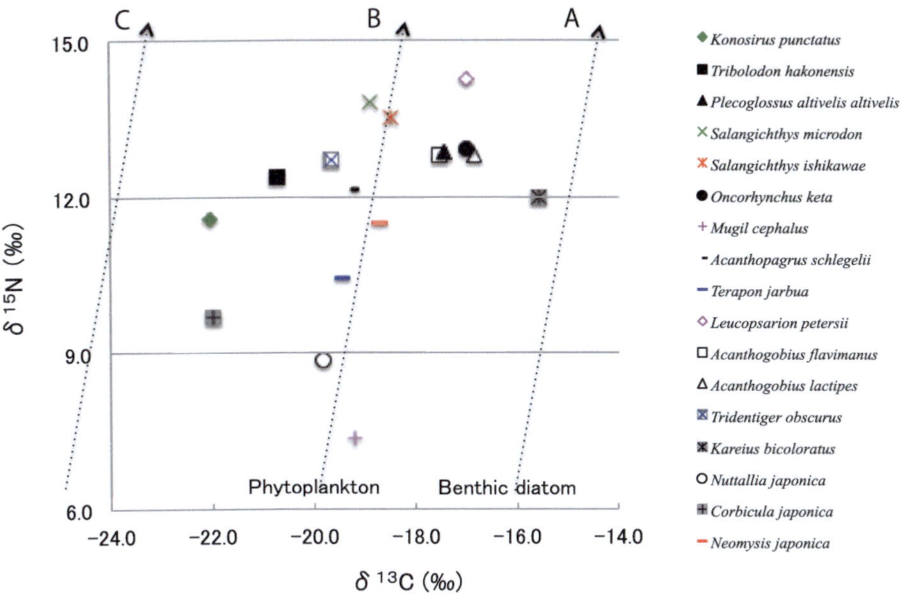

Fig. 14.10 Plots of mean $\delta^{13}C$ vs.$\delta^{15}N$ of organisms sampled at the mouth of the Natori River in April and May in 2012. *Broken lines* indicate food chains based on: *A*; benthic, *B*; pelagic, *C*; detritus

The landing of fish such as flounder in Iwate Prefecture increased after the tsunami (Goto 2014) despite only minor changes and effects seen in the fishery resources in Iwate coastal waters (Goto and Omura 2012). The tsunami caused large changes in the sea bed, moving sediment and depositing deep beds of mud but there was little change in the water column other than the temporary presence of large amounts of dissolved organic matter and suspended inorganic material. Pelagic organisms such as fishes are able to escape from badly affected areas by swimming away, but of course for many benthic animals the ability to move is limited.

While the populations of most fishes in the brackish water environment showed damage limited mostly to local and temporal effects, other marine coastal environments suffered more profound damage, Around the Oshika Peninsula, the crusted coralline area (CCA) suffered severe damage, while less damage was seen in areas of macro algal forest. Juvenile urchins and abalone inhabiting the CCA decreased to only 5 and 14 %, respectively, of their population densities just before the tsunami (Takami et al. 2013). In Gamo Lagoon, of 79 species of macrobenthos recorded before the tsunami, 47 (mostly bivalves) were completely absent or drastically reduced in numbers following the tsunami (Kanaya et al. 2012). Species absent after the tsunami included 30–80 % of taxa inhabiting the intertidal flats in Sendai Bay and the Ria coast of Sanriku, with endobenthic and sessile epibenthic forms proving to be more vulnerable to the tsunami than more mobile benthic animals like shore crabs and snails (Urabe et al. 2013). The main factors affecting survival after the tsunami seem to be habitat, life style and life stage. Intriguingly, there was little apparent damage to bivalves such as the ark shell *Anadara broughtoni* (Akagai) and surf clams *Pseudocardium sachalinensis* (Ubagai or Hokigai) inhabiting coastal areas of Sendai Bay (Watanabe et al. 2013).

These are endobenthic animals inhabiting neritic areas. It is of ongoing interest to investigate further the relationships between degree of impact and the life style of different groups of organisms.

14.4.2 The River Mouth as a Region of Largest Impact from the Tsunami

The direct death of organisms at the mouth of the Natori River, especially bivalves, was due to their being lifted off their substrate or moved out to the sea by the tsunami. Few bivalves were found inhabiting the river mouth after the tsunami, which is in stark contrast to the apparent light damage suffered by bivalves in coastal areas of Sendai Bay (such as *P. sachalinensis*, as just mentioned). In Sendai Bay, sediment analysis of C and N distributions suggests that there was little change in sediments as a whole, but areas close inshore suffered larger changes, particularly because of the deposition of large amounts of terrestrial soil transported by the receding tsunami (Gambe et al. 2014). As a result, there have been drastic morphological changes to the coast, and the mouths of rivers have been severely eroded (Tanaka et al. 2013), effects which were manifested in the studies presented here.

14.5 Conclusions

Much data is now accumulating from records and reports on the effects of the tsunami, and this must be used wisely to improve our understanding of how aquatic life responds to such disasters, and therefore how it might be possible to mitigate the effects of similar disasters in the future. Concerning fishery resources, different effects on (and recovery of) organisms were found to be related to the particular type of organism and the particular physical conditions of the environment. For very small areas such as a river mouth, it was possible to effectively monitor the different forms of life and biological production processes to obtain information about the local fishery resources with an important nursery function. The present study also emphasizes that biological production systems are continually undergoing changes. Gathering information about the relation between food organisms and environmental conditions is very important to realize sustainable production of fish resources. Our understanding can be further enhanced by investigating the ways in which changes can be brought about through both the effects of natural disasters (such as the recent tsunami) and of human impact.

Acknowledgements The studies reported here have been supported by the "Tohoku Ecosystem-Associated Marine Science (TEAMS)" project funded by the Japanese Ministry of Education, Science and Technology. We thank the Hirose Natori River Fishery Cooperative and Miyagi Prefecture Fisheries Technology Institute Freshwater Fisheries Experimental Station for help with surveys. We thank the Sendai River and National Highway Office of MLIT for provision of the information about the Natori River. We are deeply grateful to Dr. Ian G. Gleadall for useful discussions and for comments on the manuscript.

References

Deniro M, Epstein S (1978) Influence of diet on the distribution of carbon isotopes in animals. Geochim Cosmochim Acta 42:495–506

Deniro M, Epstein S (1981) Influence of diet on the distribution of nitrogen isotopes in animals. Geochim Cosmochim Acta 45:341–351

Fisheries Research Agency of Japan (2013) www.fra.affrc.go.jp/eq/repo_res/report01.pdf

France RL (1995) Carbon-13 enrichment in benthic compared to planktonic algae: food web implications. Mar Ecol Prog Ser 124:307–312

Fry B, Sherr E (1984) $\delta^{13}C$ measurements as indicators of carbon flow in marine and freshwater ecosystem. Cortrib Mar Sci 27:13–47

Gambe S, Oota H, Suzuki N, Ito K, Sasaki K, Inomata K, Nakagawa R (2014) Presumption of sediment movement in Sendai Bay caused Pacific coast of Tohoku Earthquake Tsunami, according to comparison of C, N quantity and stable isotope ratio. Miyagi Pref Rep Fish Sci 14:1–10 (in Japanese)

Geospatial Information authority of Japan (2015) Data from basic reference points and observations from geodetic surveys. http://www.gsi.go.jp/kizyunten.html (in Japanese)

Goto T (2014) Stock size fluctuations of Japanese flounder, *Paralichthys olivaceus,* in the coastal waters off Iwate Prefecture after the Tohoku earthquake and tsunami disaster. Kaiyo Mon 46(11):12–20 (in Japanese)

Goto T, Omura T (2012) Effects of earthquake disaster with huge tsunami in March 2011 on marine environment and fishery resources in the coastal waters off Iwate, Pacific coast of northern Japan. Kaiyo Mon 44(6):328–335 (in Japanese)

Honda H, Katayama S, Ito K, Chida Y, Omori M, Okata A (1997) Structure and function of the production system in estuarine fish assemblage. Bull Coast Ocean 35:57–68 (in Japanese)

Ishida O, Ishii T (1972) Tolerance to salinity of *Corbicula japonica* and morphological differences by regions. Aquac 19:167–182 (in Japanese)

Ito K (2002) Carbon and nitrogen stable isotope ratios of planktonic and benthic diatoms. Jpn J Limnol 63:166–168 (in Japanese)

Ito K, Kakegawa T (2007) Food source for fishes in blackish water ecosystem. In: Tominaga O, Takagi N (eds) Discovery in aquatic animal ecology presented by stable isotope scope. Koseisha Koseikaku, Tokyo (in Japanese)

Katayama A, Ito K (2012) Distribution of the brackish-water clam *Corbicula japonica* in the Natori River after the Great East Japan Earthquake. Kaiyo Mon 44(6):314–320 (in Japanese)

Katayama A, Ito K (2014) Recovery of resources of *Corbicula japonica* in the Natori River after the Great East Japan Earthquake. Kaiyo Mon 46(11):31–38 (in Japanese)

Katayama A, Ito K, Sasaki K, Katayama S (2013) Carbon and nitrogen sources for the bivalve *Corbicula japonica* in Natori River estimated from stable isotope analyses. Nippon Suisan Gakkaishi 79(4):649–656 (in Japanese)

Kanaya G, Suzuki T, Maki H, Nakamura Y, Miyajima Y, Kikuchi E (2012) Effects of the 2011 Tsunami on the topography, vegetation, and macro benthic fauna in Gamo Lagoon. Jpn J Benthol 67:20–32 (in Japanese)

Minagawa M, Wada E (1984) Stepwise enrichment of $\delta^{15}N$ along food chains: further evidence and the relation between $\delta^{15}N$ and animal age. Geochim Cosmochim Acta 48:1135–1140

Ministry of Land, Infrastructure, Transport and Tourism (2013) www.milt.go.jp/common/000143300.pdf

Saino T, Hattori A (1980) ^{15}N natural abundance in oceanic suspended organic particulate matter. Nature 283:752–754

Shizuka K (2013) Study on the body size variation in ayu population of the Hirose River. Unpublished master's thesis, Tohoku University (in Japanese)

Shizuka K, Ito K, Sasaki K, Katayama S, Yusa K (2012) Effects of the great earthquake and tsunami on the running upstream, growth and maturation of ayu *Plecoglossus altivelis altivelis* in the Natori and Hirose Rivers, Northeastern Japan. Kaiyo Mon 44(12):693–699 (in Japanese)

Takahashi I (2005) Study on the early life history of ayu in the Shimanto Estuary, Japan. Bull Mar Sci Fish Koch Univ 23:113–173 (in Japanese)

Takami H, Won N, Kawamura T (2013) Impacts of the 2011 mega-earthquake and tsunami on abalone *Haliotis discus hannai* and sea urchin *Strongylocentrotus nudus* populations at Oshika Peninsula, Miyagi, Japan. Fish Oceanogr 22:113–120

Tanaka H, Mohammado BA, Roh M, Vo CH (2013) Sand spit Intrusion into a river mouth after the Great East Japan Tsunami. J JSCE Ser Environ Res 69(2):616–620 (in Japanese)

Urabe J, Suzuki T, Nishita T, Makino W (2013) Immediate ecological impacts of the 2011 Tohoku Earthquake Tsunami on intertidal flat communities. PLoS ONE 8:e62779. www.plosone.org

Watanabe K, Shoji M, Sasaki K (2013) Impact on the bivalve Fisheries in the middle-south area of Sendai Bay brought by the Great East Japan Earthquake. Miyagi Pref Res Fish Sci 13:23–29 (in Japanese)

Index

V. Santiago-Fandiño et al. (eds.), *Tsunamis and Earthquakes in Coastal Environments*, Coastal Research Library 14, DOI 10.1007/978-3-319-28528-3

MIX
Papier aus verantwortungsvollen Quellen
Paper from responsible sources
FSC® C105338

If you have any concerns about our products,
you can contact us on
ProductSafety@springernature.com

In case Publisher is established outside the EU,
the EU authorized representative is:
Springer Nature Customer Service Center GmbH
Europaplatz 3, 69115 Heidelberg, Germany

Printed by Libri Plureos GmbH
in Hamburg, Germany